Documentarity

History and Foundations of Information Science

Edited by Michael Buckland, Jonathan Furner, and Markus Krajewski

Human Information Retrieval by Julian Warner

Good Faith Collaboration: The Culture of Wikipedia by Joseph Michael Reagle Jr.

Paper Machines: About Cards & Catalogs, 1548–1929 by Markus Krajewski, translated by Peter Krapp

Information and Intrigue: From Index Cards to Dewey Decimals to Alger Hiss by Colin B. Burke

Indexing it All: The Subject in the Age of Documentation, Information, and Data by Ronald E. Day

Bibliometrics and Research Evaluation: The Good, the Bad, and the Ugly by Yves Gingras

Search Foundations: Toward a Science of Technology-Mediated Experience by Sachi Arafat and Elham Ashoori

The Information Manifold: A New Perspective on Algorithmic Bias, Fake News, and the False Hope That Computers Can Solve These Problems by Antonio Badia

Documentarity: Evidence, Ontology, and Inscription by Ronald E. Day

Documentarity

Evidence, Ontology, and Inscription

Ronald E. Day

The MIT Press
Cambridge, Massachusetts
London, England

© 2019 Massachusetts Institute of Technology

All rights reserved. No part of this book may be reproduced in any form by any electronic or mechanical means (including photocopying, recording, or information storage and retrieval) without permission in writing from the publisher.

This book was set in Stone Serif and Stone Sans by Jen Jackowitz. Printed and bound in the United States of America.

Publication of this open monograph was the result of Indiana University's participation in TOME (Toward an Open Monograph Ecosystem), a collaboration of the Association of American Universities, the Association of University Presses, and the Association of Research Libraries. TOME aims to expand the reach of long-form humanities and social science scholarship including digital scholarship. Additionally, the program looks to ensure the sustainability of university press monograph publishing by supporting the highest quality scholarship and promoting a new ecology of scholarly publishing in which authors' institutions bear the publication costs.

Funding from Indiana University made it possible to open this publication to the world.

Library of Congress Cataloging-in-Publication Data is available.

ISBN: 978-0-262-04320-5

10 9 8 7 6 5 4 3 2 1

Of would be a difficult person to like.
—Carla Harryman, 1992

There are no fungibles in nature.
—Rom Harré, 2018

Contents

Acknowledgments ix

Introduction 1

1 **The Philosophical Perspective** 11

2 **Documentarity in the Works of Paul Otlet and Georges Bataille: Two Competing Notions of "Document" and Evidence** 37

3 **Figuring Documentarity** 51

4 **Documentarity and the Modern Genre of "Literature"** 65

5 **Displaced Reference for Information: Jokes, Trauma, and Fables** 99

6 **Rights of Expression** 111

7 **Post-Documentation Technologies** 137

Conclusion 151

Notes 153
References 161
Index 169

Acknowledgments

I am very grateful to my editor at the MIT Press, Gita Devi Manaktala, for her support of this book, her patience with its extended writing time, and her superb choice of reviewers. And I am grateful for the anonymous reviewers of this book for their work; their insights and labor greatly assisted me over several revisions. Without them, this book would have been much less. I'd like to acknowledge Michael Buckland, Maurizio Ferraris, Geoffrey Bowker, and Johanna Drucker for their many years of research works and for conversations with them, which have greatly contributed to this project. In Michael's case, I'd like to also thank him for reading the manuscript in various stages, as well. I'd like to thank Neal Thomas for his insightful and learned works, which have helped me better understand "post-documentation" technologies. I'd like to thank Clovis Ricardo Montenegro De Lima and Marcia Heloisa Tavares de Figueredo Lima for hosting me in Rio de Janeiro during several visits, which led to my discussion of the right to truth in chapter 6. I'd like to thank my family, Jiangmei Wu and Dexter Wu-Corts, and my friend Claire McInerney, for their personal support during the writing of this book, Dr. Scott Breeden for his friendship and keeping me fit during its composition, and Reine Melvin for her friendship and support while I researched Suzanne Briet's works in Paris. The writing of this book was supported by small grants from the Association for Information Science and Technology's (ASIS&T) History and Foundations of Information Science special interest group and from the Rob Kling Center for Social Informatics at Indiana University at Bloomington, for which I am grateful. This work was (partially) funded by the Office of the Vice Provost of Research and the IU Libraries. I'd like to thank the series editors, of which this book is a part, Michael Buckland, Jonathan Furner, and Markus Krajewski, for all their work and consideration of this text in their series. I'd like to thank Judith M. Feldmann of the MIT Press and Michael Goldstein for their production work and copyediting of this book. I'd like to thank the librarians at Indiana University at Bloomington libraries for the absolutely wonderful library they maintain, without which I could not have researched this book. And last, I would like to thank the readers of this book for taking time out of their lives in order to consider what I have written.

Introduction

I

This book crosses philosophical ontology and documentary ontology, trying to understand different genres, technologies, modes of inscription, and innate powers of expression by which something comes to appearance as what is. Instead of approaching information ontologically ("what is information?") it approaches ontology informationally, asking how something becomes evident and is taken as evidence of what is. It is a story of some of the inscriptional means for this, which will become clear by the end of the book, are not only information and knowledge technologies, but in the last resort, also *technologies of judgment*. What are judged are entities as to their identities, qualities, and actions, not just descriptively, but prescriptively.

I call the philosophy of evidence "documentarity," and I am interested in how this philosophy of evidence constitutes Western (cultural and disciplinary) philosophy as the Western metaphysical tradition,[1] largely understood as a representational tradition in its epistemology, ontology, aesthetics, and politics. Documentarity is the philosophy of what comes into presence and makes itself evident, foremost in representation. The purpose of this book is to describe documentary evidence, and so, documentation, within a tradition of Western metaphysics, and so, largely, representation (though to do so I have traced the outer limits and contrary epistemologies and ontologies of such, as well). The study of documentarity is the study of technologies of judgment at the level of ontology—in this case, what and how what is appears to us as such.

This book discusses the expression of beings or entities as evidence of what is, from ancient categories to medieval figures to modern documentation and modern literature to sublime and dispositional expressive powers, and finally, to social networks and machine learning. Documentarity is the philosophical basis of the practice and theory of documentation; it is traditionally understood as "representational" in various manners. Documentation is the foundation for much of what we call "information," today, and its philosophical commitments shape our understanding of all manners of expression, from institutional documents to published opinion and even to literature and the other arts, too. Seen from a contemporary perspective, therefore, this is a book about information as truth and how this evolves within a tradition—documentation and its originary philosophy of documentarity—where truth is seen in the appearance of what is, through inscription, and foremost through, and as, representation.

As I will show, the range of types, genres, and modes of evidence for what is varies across different discursive fields and seems to follow a general historical trajectory from Ancient works until today along a course from ideal reference (which I call "strong documentarity") to phenomenological and linguistic sense (which I call "weak documentarity"). The transition from strong documentarity to weak documentarity traces the greater role that semiotic and empirical senses are given to creating and stabilizing reference. In strong documentarity, reference is seen as a product of categories (as in Plato's philosophy, or in traditional documentation techniques). In weak documentarity (which may end up being strongly representational, however), empirical particulars play a stronger role, either as innate dispositional powers or through empirical research. This is a more Aristotelian perspective, where categories of being are results of descriptions of the expressive powers of particulars.

I hope that this book shows that there are not only theoretical concerns that belong to a philosophy of evidence, but also practical and political concerns and aesthetic forms of such. Documentarity and, today, "information," have not only epistemic, but also political and aesthetic, qualities that should be taken account of. The philosophy of evidence not only governs knowledge institutions, but has also governed political institutions and has managed the role of knowledge in colonialism and the spread of "the West" and its notions of literacy and their importance. It has also colored the notion of what modern literature is (and more generally, the arts are) as realism and sometimes as formalism. "Information," understood as a substantive, also has an aesthetic form, namely of organic representation.

Too little attention has been paid to the aesthetics of information, and how this involves epistemic and practical concerns.

II

This is my third book on information as metaphysics. In *The Modern Invention of Information: Discourse, History, and Power* (2001), I looked at the then contemporary conception of information, within its twentieth-century Western European and North American contexts, as a trope for the metaphysical notion of "presence" (using this term in a Derridean sense). In *Indexing It All: The Subject in the Age of Documentation, Information, and Data* (2014), I examined information according to the information-seeking and retrieval paradigm in information science: persons and documents as dialectically constructed subjects and objects—what I called the *modern documentary tradition*. This present book takes the story of the metaphysics of information back to ancient Greek philosophy and encompasses the Western metaphysical tradition more broadly than the previous books.

This book, however, continues a meditation upon the meaning of the figure of the antelope that begins Suzanne Briet's 1951 book *Qu'est-ce que la documentation?* (translated in 2006 as *What Is Documentation?*). It is an attempt to theoretically think the particular antelope both before and after the processes of its capture and being made evidence by means of traditional documentation techniques and institutions. I asked myself: What is forgotten when a particular being—an antelope—is understood as being a universal type (a species type), that is, when an entity is read in terms of its class "essence" (e.g., as a new type of antelope)? What is forgotten about particular beings when they are subject to (or subjects of) the representation of being, understood as essential universal types (i.e., as class members)? How do such class types appear through *a priori* and through *a posteriori* methods, and through different genres of writing? How have we thought of beings in the Western tradition as evidence of truth, and how have beings been made to conform to this by the theoretical and practical powers of inscriptions and documentary institutions? How do entities represent themselves as evidence?[2] And, also, how does the category of "literature" or art evolve in modernity as both a reaction to "factual" documentation and also as an intensification of such?

This last question itself refers back, as well, to the very mixed genre of Briet's *Qu'est-ce que la documentation?* as both an historical-literary manifesto for documentation and as a technical manual. In Briet's book, this shows in the formal construction of the book: the very persuasive and literary first part, and the more technical second and the even more technical third parts. In Briet's total oeuvre, however, this question also conversely appears in the starker division between her professional works on documentation and her works on the oeuvre of the poet Arthur Rimbaud, both without any traces of the other. In this present book, I cross these genres in order to show how traditional documentation, literature, and art intersect within the metaphysics of documentarity.

III

This book has the topology of a "U" in its duration. Overall, though with shifts and changes within the chapters and overviews in the first chapter, it begins with a discussion of entities as belonging within types or classes of representation and it progresses to discussing them as inscriptions and dispositional powers, eventually being represented through predictive algorithms in machine learning. The middle of the book discusses the works of the avant-garde narrative writer Carla Harryman and the poet Barrett Watten, and these constitute a sort of "near-zero" point in representation, where, following the modern avant-garde tradition, techniques of inscription come to the fore.

IV

For some of the more complex chapters, I start with an italicized leading sentence in order to suggest to the reader what is contained therein.

The chapters of this book may be summarized as follows:

Chapter 1: The Philosophical Perspective

This chapter addresses the topic of the book in terms of the Western philosophical tradition, reaching back to Plato's understandings of entities in terms of their being evidence of some class or idea, and for Aristotle, as their being essential sets of expressive powers. The chapter begins with Martin

Introduction

Heidegger's return to these issues in terms of the question of being and of inscription (*poiesis*), including the inscriptionality of *techne* and modern technology. It then continues in providing an overview of the entire contents of this book from a philosophical perspective, reaching forward in time to Bruno Latour's notion of inscriptions and ontological modalities, and then to Rom Harré's epistemology of dispositions and affordances as the basis for an expressionist theory of entities, which we will return to at the end of the book.

Chapter 2: Documentarity in the Works of Paul Otlet and Georges Bataille: Two Competing Notions of "Document"

This chapter examines opposed metaphysical notions of what documents are and do in the works of the early twentieth-century thinkers and librarians, Paul Otlet and Georges Bataille, the first of a positivist origin and the other of a materialist origin that is grounded, in part, in early twentieth-century French ethnology and the modern artistic and literary avant-garde. We see here a contrast in notions of evidence, collections, and truth between a positivist documentation tradition and a modern anthropological one grounded in performance and experience. The difference inscribes two views of documentarity in regard to documents: representational metaphysics and a Nietzschean type of overturning of such. However, I argue that this particular "overturning" of ideal *reference* by phenomenological *sense* is also metaphysical.

Chapter 3: Figuring Documentarity

This chapter begins by engaging John A. Walsh's (2012) work on the documentary index as a process of figuration in medieval iconography and allegory. The chapter then returns to Suzanne Briet's *What Is Documentation?* (1951, 2006), to examine the ontological and epistemic importance of the documentary index in what she sees as modern science. We turn to the medieval tradition in order to examine art and literature as genres, where the documentary index has metonymical and allegorical figuration, in order to demonstrate an experiential indexicality, though one grounded in theological semiotics. Much later, Briet's documentation theory prioritizes scientific explanations, but this epistemology of science is one grounded in bibliographic traditions of classification and ontology, and so I will argue that it, too, is colored in a semiotic manner with revelatory figuration playing a major role in both the theory and the practice of naming an entity as evidence of a genus or species of being. As we will see more broadly throughout the book, documentarity has both static (classification) and revelatory (algorithmic) forms for making evident, blocking out competing psychologies of time and experience.

Chapter 4: Documentarity and the Modern Category of "Literature"
This chapter examines modern literature and art in two cases: first, a dialectical relation to strong documentarity in nineteenth-century French Realism; and second, the twentieth and into the twenty-first-century literary and aesthetic avant-garde's critique of representation. The nineteenth-century French realist novel genre critiqued, but also intensified, documentary realism by creating characters with intense and subtle emotional range and internal mental dialogue, situated within, and responsive to, modern social conditions (not least, created by mass media). Like today's "information," it collapsed the space between the reader and itself as document, inscribing the reader in the world of its own senses, leaving the reader to enact interpretive reading in the analogical application of its "content." Reference was intensified by means of complexities of sense, both in regard to the internal states of the characters and the world around them. In contrast, the modern avant-garde, particularly in the early twentieth century, pushed toward a counterform of evidence rooted in sensation and in the creation of new forms out of semantic materials. It aimed toward a reality distinct from documentary realism.

Chapter 5: Displaced Reference for Information: Jokes, Trauma, and Fables
In this chapter, we look at performative genres where the expression of powerful particulars is indirect and mediated by literary devices. Evidence appears through representational performances, but those performances are means for human powerful particulars themselves to appear as witty or as traumatized or as actualized meaning. The "I" appears as a powerful particular, not representationally framed, but as inflected through, performative literary devices and forms. The empirical particular is exemplified, but still through mediating literary devices (even in psychological diagnosis and therapeutics, per historiographic frames and devices).

Chapter 6: Rights of Expression

This chapter examines self-evidence by expressive particulars, stretching from humans to ecological bodies, each read in term of the ontology of Rom Harré's notion of "powerful particulars" (Harré, 1995). The chapter examines such an ontology in modern rights discourse, particularly in the "rights drift" from natural rights to the international law principle of the right to truth to animal rights and then to rights of nature. This chapter views particular real entities as inherently ontologically expressive and with organic bodies having rights by virtue of their unique dispositional powers (rather than as, in the documentation tradition, having such by virtue of the representational classes or descriptive vocabulary to which they are assigned through ontologies and other cultural practices of identification, attribution, and relationship). Though mediated and expressed by nominal social and cultural affordances that shape the nature of rights and expressions, organic real entities are unique entities of powers that can assemble such nominal tools as tools of expression. They fold semiotic materials into dispositional tools, and they are singular no matter how much they "belong" to classes. Their "freedom" or "will" is both socioculturally constituted in experience and unique and radically singular by virtue of their developmental learning. While all entities, both organic and inorganic are unique at some level of their material being, rights are generally given to those whose uniqueness is displayed in ordinary human experiential worlds. "Freedom" and "will" are assigned to those entities and (and even to those classes of entities) that "push back" upon human *a priori* classification and standardized descriptions (e.g., the broadest class of that which pushes back upon classification is the class of entities that we call the natural world or nature).

Chapter 7: Post-Documentation Technologies

This chapter introduces the notion of "post-documentation" technologies, which include social media technologies and machine learning. Post-documentation technologies still remain part of the paradigm of documentarity, but they are largely *a posteriori*, rather than *a priori*, techniques and technologies for calculating sensible tendencies and directionalities for entities. (*A priori* techniques are used as parameters, initial ontologies, or heuristics, and appear as classes as the result of processing.) The "post" of "post-documentation," signifies a break from traditional documentation that, however, doesn't escape the larger general economy of documentarity. (I have used "post-documentary" or similar terms, rather than "post-documentation," in previous writings, but for the purposes of this book I have chosen "post-documentation.") The parameters that reorganize empirical traces as essential qualities or predictive trends are not only technological, but also socioculturally "ideological" (in the general, as well as the politically specific, way that I will use this term in this book, namely as constellations or assemblages of ideas). Whereas older technologies of documentation, such as classification, had obvious and explicit *a priori* technologies for producing evidence, newer post-documentation technologies tend to function in real-time and as infrastructural technologies. Post-documentation technologies are what I call *computer-mediated judgments*. Computers, particularly as personally and socially embedded and mediating devices, are very powerful *technologies of judgment*.

IV

I will conclude this introduction by mentioning that every book has precedent works before it that have influenced it. For this book, the initial work that played this role was Erich Auerbach's *Mimesis: The Representation of Reality in Western Literature*. From this perspective, this present book attempts to address inscriptional (broadly understood, "literary") representations in terms of evidence and documentation.

This book also attempts to address some of the theoretical issues of modal expression that I felt were incompletely analyzed in Bruno Latour's *An Inquiry into Modes of Existence*, most of all by contributing an epistemology of dispositions and affordances to the problem of explaining modalities of expression.

In addition, Maurizio Ferraris's book, *Documentality: Why It Is Necessary to Leave Traces*, also provided a gentle antagonist and a model in some aspects for this book. In his book, Ferraris understands documentality in terms of textuality, a relationship that I reverse in this one, giving metaphysics as a technology of evidence and evidence production a stronger role in structuring textuality. I have coined the term "documentarity" to distinguish this term and the approach of this book from Ferraris's notion of documentality, giving a stronger American "r" to the lightness of Ferraris's textual "documentality," as it were. Following from my earlier book, *Indexing It All*, I view information technologies as indexing together documents and concepts with subjective agencies into being identities, attributes, dispositions, and powers, in other words, into the qualities and modalities of being. Through digital mediation, this being often takes the form of inscriptions, information flows, conduits of action, and recognitions and self-recognitions. I am interested in this book in the role of representational frames and devices in constructing these and their genres and modes of textuality.

And as will be obvious to the reader, I am also greatly indebted to Rom Harré's many works, his explanations of dispositional powers and affordances in many disciplines, and his attention to both inscription and the ontological powers of entities, from quantum particles to human beings.

Last, this book, like many of my works, contemplates the status of what it means to be a being in an age of evidentiary, particularly today digitally mediated, inscription; specifically, as a human being, of course, but also as any entity with powers of expression. And so, this work remains tied to the problematics of *techne* and *poiesis* in Martin Heidegger's works, to dispositional models of expression in Rom Harré's works, and as mentioned earlier and analyzed in my earlier books, to the problem of the relationship of technique and being in establishing identity, meaning, value, and truth for entities.

Each entity in the world is singular at some level of its ontology (Harré, 2018), in so far as its development is guided by uniquely historical and situational developments, and with higher-level organisms, learning from experiences and developing personal toolkits of expression from these. Each entity has intrinsic and, with organic beings, inherited or innate (from evolution and from learning) dispositions. So, we may ask, through what powers does an entity come to appear to us as being? Through what technologies of judgment and thought, and the enfolding of these in ancient and modern inscriptional machines (and the further forming of judgment and thought from these) does an entity become *evident* to us? What is the historical appearance and the historicity of such judgments, thoughts, and their inscriptional machines, and how do they manifest today as horizons for our and others' judgments and evidence? These are the questions that guide this book.

1 The Philosophical Perspective

The Idea of an Idea Is a Rather Peculiar Idea

In this chapter, I will present the outline of various philosophical discourses that conceptually and historically frame the entirety of this book. The chapter begins with Martin Heidegger's engagement with the question of being at what he saw as the historical and conceptual ends of Western philosophy. It then discusses Bruno Latour's notion of inscription. And it ends with presenting a reading of Rom Harré's theory of dispositions and affordances, which will give us the tools to better understand the expressive powers of particulars. These themes will later be explored in the book through classificatory, indexical, descriptive, performative, ironic or "unconscious," self-expressive, and finally, social and predictive rhetoric and technologies of evidence production, along a conceptual and historical horizon, leading from *a priori* to *a posteriori* or "empirical" modes of evidence production.

Martin Heidegger: *Poiesis* and the Task of Thinking

In his late lecture of 1966 (delivered in his absence in France), entitled in English as "The End of Philosophy and the Task of Thinking," Martin Heidegger opened with a critique of his work from nearly forty years earlier, *Being and Time*:

> The title ["The End of Philosophy and the Task of Thinking"] designates the attempt at a reflection that persists in questioning. Questions are paths toward an answer. If the answer could be given it would consist in a transformation of thinking, not in a propositional statement about a matter at stake.
>
> The following text belongs to a larger context. It is the attempt undertaken again and again ever since 1930 to shape the question of *Being and Time* in a more

primordial fashion. This means to subject the point of departure of the question in *Being and Time* to an immanent criticism. Thus, it must become clear to what extent the *critical* question, of what the matter of thinking is, necessarily and continually belongs to thinking. Accordingly, the name of the task of *Being and Time* will change.

We are asking:

1. What does it mean that philosophy in the present age has entered its final stage?
2. What task is reserved for thinking at the end of philosophy?

(Heidegger, 1977b, p. 373)[1]

For Heidegger, the critical question, the question of the critique of metaphysics, is that of trying to think the question of being at the end—the culmination—of philosophy. Philosophy, for Heidegger, means the Western metaphysical tradition, which is guided by ontology, the study of being. Further, as the text makes explicit by a reference to "thinking," "philosophy" for Heidegger doesn't just mean academic philosophy, but rather, it means, at least in addition to this, a culturally and socially specific understanding of the fundamental question of *Being and Time*, namely, what is and how such comes to present itself to us. As such, Heidegger is engaged in a critique of the foundations of Western metaphysics through ontology, and with that, problems of identity and the appearance of evidence. The Western philosophical tradition for Heidegger presents in microcosm the cultural tradition of "the West," from its beginnings in Plato and Aristotle through its practical unfolding in science and technological modernity. And in so far as questioning is, for Heidegger, always a path, the critical questioning of being itself marks a way along a path, now at an "end of philosophy"—what we could translate from the German philosopher as not only an arrival as to what philosophy is in the Western tradition, but also as a "dead" or "defunct" end, not because it comes to a stop, but because, as Heidegger writes,

> The end of philosophy is the place, that place in which the whole of philosophy's history is gathered in its most extreme possibility. End as completion means this gathering. (1977b, p. 375)

For Heidegger, philosophy is not only in the Western tradition, but the Western tradition—and the very concept of such—*is* philosophical, not only theoretically, but also practically in its social and cultural unfolding on a global scale. Philosophy is the logic of power in and as "the West," culminating in global, technologically driven modernity. Significantly, Heidegger adds:

> The end of philosophy proves to be the triumph of the manipulable arrangement of a scientific-technological world and of the social order proper to this world. The end of philosophy means the beginning of the world civilization based upon Western European thinking. (1977b, p. 375)

For Heidegger, modernity is the practical fulfillment of Western metaphysics as a philosophy of ontological presence. Presence in metaphysical ontology is known representationally, through ideas and paradigms. For Plato, ideas are essential and universal types that are expressed in evidential particulars, and paradigms (e.g., in *Phaedrus* and *The Sophist*), are forms that gather many examples together so we may know examples as examples of some universal form or idea (*eidos*).[2] As I will discuss in this book, "empirical data," too, gathered through paradigms, is constrained by ideas, either as *a priori* or as functional *a posteriori* parameters for thinking. This is the very nature of representational thought; it is an ideational and paradigmatic aesthetic. What is present is contained and guided into appearance by representation and representational processes.

Why does Heidegger write that we must "subject the point of departure of the question in *Being and Time* to an immanent criticism"? This is a very important question, not only in regard to this lecture of Heidegger's, but moreover in regard to the entire Heideggerian project, as it was announced in *Being and Time*.

In the "End of Philosophy and the Task of Thinking," Heidegger is questioning his philosophical project since *Being and Time*. He is searching for a fundamental ontology in the midst of a critique of understanding the being of beings in terms of representation and technologically driven reproduction. Within such a critique, the notions of *poiesis* and *techne* will play a major role.

The German of our beginning quote reads: "Es ist der seit 1930 immer wieder unternommene Versuch, die Fragestellung von *Sein und Zeit* anfänglicher zu gestalten. Dies bedeutet: den Ansatz der Frage in *Sein und Zeit* einer immanenten Kritik zu unterwerfen." Here Heidegger writes that since 1930, he has attempted to reformulate the *"Fragestellung"*—the positioning of the central question—of *Being and Time*. This means to subject the starting approach (*"Ansatz,"* in which one hears "Satz" or proposition [the English here reinforces the issue of the *Stellung* or positioning of such an approach]) to critique. Here, critique plays the role of a reoccurring intervention into the very paradigm that characterizes it: an intervention into the *assertion* or statement of fact—the propositional re-presentation—of entities and events. It is an intervention into the asserted form of representing entities and events as statements of fact. This intervention had earlier started with Heidegger's investigations into *poiesis* and its relationship to philosophy as knowledge propositions about the world and as the imagination of the world itself as a knowable object.

Returning to Husserl's "call to the thing itself," Heidegger writes, "Für den Ruf ist nicht die Sache als solche das Strittige, sondern ihre Darstellung, durch die sie selbst gegenwärtig wird" ("It is not the matter which itself is controversial for the call, but rather its representation, through which it itself becomes present").

Here, I have modified the Krell translation of the 1966 lecture that I have been using, because *Darstellung* is an issue not just of presentation, but of representation. The very notion of modern science that Heidegger criticizes here is that of science as a notion of "enframing"—of representations by which something is brought to presence as an object for subjectivity. Instead, the problem of presence or "presencing" is for Heidegger an issue of different manners of *poiesis* and the techniques and technologies (*techne*) through which *poiesis* occurs. For this reason, critique so often occurs in Heidegger's later works through a contrast of rhetorical genres for presenting what is, including in the rhetorical performances of his own writings on philosophy.

How did Heidegger see the problem of the representation of "the thing itself" play out in the culture of modernity, so that the initial project of *Being and Time* had at least to be redirected in some way beyond the initial philosophical approach of that book?

Foremost, Heidegger saw the culmination of metaphysics in the dominance of the *Gestell* of the representational worldview. For Heidegger, philosophy (what we could call a "cultural metaphysics") pervades the "world picture" of modernity, beginning with the very concept of "the world," understood as an object distinct from daily existence. Heidegger's notion of "enframing" (*Gestell*) is one based on the representation of entities as essences whose properties are transcendental to inscriptional affordances for expression. This is accomplished by narrowing or "enframing" the entity only in terms of certain affordances so that only certain dispositions appear, namely those that serve technological means to ends within chains of resource exploitation. These dispositions are then taken themselves as initial and final causes—essences—of entities. Transcendental philosophy arrives at its "truths" through practical, anthropocentric means. For Heidegger, this is not the essence of scientific research, but rather of modern technology. In the modern period, however, science often is technological, not just in its tools, but in its research, becoming engineering for human ends.

Earlier than "The End of Philosophy and the Task of Thinking," Heidegger had argued that such characterized entities are taken in modern technology as "standing reserves"—*Bestand*—resources to satisfy particular human needs of use (the "will") and the technological economies that exist to serve them (Heidegger, 1977c). For example, a field that has coal is taken solely as a coalfield and is mined so, leaving devastation to the field and leaving polluted runoff.

Modern technology, for Heidegger, brings something into presence in such a way so that the entity appears as fixed in its essence. This fixity, this documentarity, this "evidence" of the thing in terms of its being a resource for energy power, wealth power, and social power, erases both the coming-to-presence of the entity by its own powers and the coming-to-presence of the entity by the technologies of inscription that are brought to it as affordances. All entities, from atoms to the world itself understood as an entity, are seen in the modern worldview as resources for human will. For Heidegger, the end of philosophy is in some ways shown by our painting ourselves into the corner of global engineering. Global engineering (today, climate geoengineering, perhaps) is not just one option, but the only logical outcome of the drive (i.e., the will to power) of metaphysical ontology when taken to its theoretical and practical ends.

What is obscured in modern technology, Heidegger argues, is the inscriptionality of philosophical metaphysics upon entities, the metaphysics of representation, which allows modern technology to act differently than previous, nonreproductive, and so unevenly systematic *techne*. Here Heidegger applies Nietzsche's critique of morality to systems science: viewing the earth and human societies as controllable "natural systems" or "worlds" is an extension of an exploitative will to power, now aided by mechanical forms which themselves act in "objective" or nonvarying manners.

An engineering perspective comes to pervade science. And that engineering perspective is ultimately tied to personal and human interests of accumulation through the exploitation of entities as things that are seen as objectively available for systemic use. For Heidegger, technoscience is often not research for research's sake, but rather research as a type of knowing that brings things into being for their systemic and repetitive use for the ends of man. All expressions of entities are taken as representational information about them for the purposes of their being engineered as resources. The information that entities express themselves as such is framed as propositional statements of what is in the mode of what is seen as essential about entities for human use. Such statements are then organized into systems of production, and eventually into self-governed, that is, cybernetic, engineering systems.

Heidegger writes in "The End of Philosophy and the Task of Thinking":

> No prophecy is necessary to recognize that the sciences now establishing themselves will soon be determined and steered by the new fundamental science which is called cybernetics.
>
> The science corresponds to the determination of man as an acting social being. For it is the theory of the steering of the possible planning and arrangement of human labor. Cybernetics transforms language into an exchange of news. The arts become regulated-regulating instruments of information.
>
> The development of philosophy into the independent sciences which, however, interdependently communicate among themselves ever more markedly, is the legitimate completion of philosophy. Philosophy is ending in the present age. It has found its place in the scientific attitude of socially active humanity. But the fundamental characteristic of this scientific attitude is its cybernetic, that is, technological character. The need to ask about modern technology is presumably dying out to the extent that technology more definitely characterizes and regulates the appearance of the totality of the world and the position of man in it. (1977b, p. 376)

Heidegger then returns to the topic of cybernetics in his conclusion:

> Perhaps there is a thinking which is more sober-minded than the incessant frenzy of rationalization, the intoxicating quality of cybernetics. One which might aver that it is precisely this intoxication that is extremely irrational. (1977b, p. 391)

Cybernetics in the post–World War II era was a transdisciplinary study of communicative feedback and control in systems, culminating in autopoietic systems. Language, too, was understood as a communicative feedback system. Norbert Wiener famously described cybernetics in 1948 as "the scientific study of control and communication in the animal and the machine" (Wiener, 1961). The attempt of cybernetics was to find universal principles for systems dynamics and their control. Such "systems" were those of human and natural organisms, as well as physical systems and mechanical systems, combined. Wiener saw such principles as centrally important for social and political governance, including what he saw as democratic governance, and he saw a great political need to formulate language as clear and distinct statements of facts that could serve the social engineering tasks of modern states.[3]

With all that is at stake in a technological worldview, why would Heidegger focus upon what he saw as the abomination of art as information exchange, that is, art as cybernetics, and why would this be seen as an exemplar of the culmination of metaphysics?

To understand this and see its relationship to documentary, we need to return to Heidegger's critique of language, understood as the exchange, or "transmission," of statements—that is, language understood as information and within information theory. This is complemented by his promotion of poetry in the war years and after as a critique against language understood in this manner:

> Within Framing, speaking turns into information [*Das so gestellte Sprechen wird zur Information*]. It informs itself about itself in order to safeguard its own procedures by information theories. Framing—the nature of modern technology holding sway in all directions—commandeers for its purposes a formalized language, the kind of communication which "informs" man uniformly, that is, gives him the form in which he is fitted into the technological-calculative universe, and gradually abandons "natural language." . . . Formalization, the calculated availability of Saying [i.e., the formalization of poetic language], is the goal and the norm. . . . Information theory conceives of the natural aspect of language as a lack of formalization. (Heidegger, 1971, p. 132)

For Heidegger, cybernetics, as an information theory based on the regulation of social systems through feedback loops of statements about entities and events, constitutes the contemporary "world-picture" (Heidegger, 1977a). The "world-picture" is not only a picture of the world, but also the world understood and mediated by information theory. The transformation of poetics or art into systems of statements of information exchange constitutes for Heidegger the culmination of the metaphysical worldview. Here, language itself becomes a fact for engineering, rather than a mode of engineering's inscription. And so engineering escapes its own existence as inscription and becomes the transcendental principle for all inscriptions.

All of this was anathema for Heidegger, for whom language was essentially poetic. By "poetic," Heidegger meant an emergent process whose inscriptionality or *techne* of beings is evident in its making (*poiesis*).

In his 1954 publication "The Question Concerning Technology," Heidegger explicates the ancient Greek concept of *poiesis* through a reinterpretation of what he claims is a later Latin philosophy reading of Aristotle's "four causes" (in Aristotle's *Metaphysics*). Heidegger replaces a deterministic reading of the four causes in consequential relation to one another (from an idea to a finished product, passing through material and labor causes) by a notion of four mutual causal affordances (*aition*) that co-afford the appearance of a natural or artistic entity. For Heidegger, causes (*aition*) hang together with one another in order to afford the emergence of an entity. Using Heidegger's example, a chalice emerges into appearance through an interdependent relationship of cultural forms (formal cause), social purpose (the final cause), physical materials (material cause), and attentive workmanship to the foregoing "causes" (efficient cause). Heidegger characterizes the poetic, or more generally, "art" (in both the fine and crafts sense of the term), according to a theory of emergence based on mutually dependent causal powers (understood as what we could characterize as affordances), while distinguishing natural from human *poiesis* by the role of human intent and human agency in the emergence or coming-forth of the object (e.g., a blossom versus a work of art, where the former has its own, innate, dispositional powers).

Heidegger reinterprets the Latin notion of "cause" (*causa*) by what he claims to be the original meaning of Aristotle's *aition*. For Heidegger, the Latin philosophy misinterpretation of *aition* by a teleological notion of *causa* prefigures later manners of seeing and exploiting the world as essential forms for instrumental ends. If all things begin with initial ideas of their essence that are brought into being through materials and energy, then their final realization, their existence as a final "cause," can be extended to serve human needs once that process is understood. Scientific research into what is becomes aimed toward the manipulation of these "natural" causal processes toward human ends. In other words, science serves engineering and engineering serves the desire to control all natural processes, now understood as resources or as threats to human will. This type of scientific rationality is an extension of an imagination of the human will as transcendental over nature, including over human nature (and its life and death). The metaphysical dream of such begins with the notion that entities and events have essential forms or "ideas" which are knowable, and that knowledge is constituted by the collection of such forms. In short, knowledge is seen as a sort of library of representational documents that record facts of nature and society.

From this perspective, the "immanent" nature of Heidegger's critique of the question of being lies in the turn from a search for a more fundamental ontology in *Being and Time* toward ontology understood as *poiesis*.[4] This is what has been termed Heidegger's *Kehre*, or "turn" (toward language), in the works following *Being and Time*. Consequently, it is not scientific research that is the object of critique in Heidegger's later work, but rather science as a mode of systems engineering based on the description of the world by essential ontological statements.

The type of naming taking place in the poetic must, for Heidegger, be separated from the type of naming occurring in documentation *qua* information, and information understood as part of a systems-engineering perspective. First because of the non-teleological process of causation that Heidegger associates with non-technological *poiesis*, and second because of the non-"methodological" or technological means toward it. (Heidegger's notion of "method"—i.e., the "thesis of the precedence of method" in "The End of Philosophy and the Task of Thinking"—is that of a strict process in fulfillment of a founding frame of inquiry; that is, it is the method of teleological causation.) For Heidegger, technoscience is characterized by the operationalization of entities and conditions so that the objects of inquiry and their context fit the founding frame and prove the hypothesis correct or incorrect. As method becomes more and more integrated into technical operations, so Heidegger claims, this operationalization becomes quicker and more efficient, erasing the questioning of *a priori* or conceptual assumptions from the science being done, as well as from a notion of science itself, seen as a free inquiry into the nature of *poiesis* in natural entities and human artifacts. (An example of this is the absence of foundational critiques in scientific articles—often the very rhetorical composition of such articles as "scientific" articles preclude such. Instead, the literature review "builds on" previous research.) Science becomes method, as the founding frames for science—its ontologies, and itself as a theory of knowledge of being—disappear as unexpressed premises in its research practices and writing.

To summarize: broadly viewed, Heidegger's critique emerges from his seeking a more fundamental ontology of things themselves via phenomenology in *Being and Time,* and then increasingly in his works after the "turn" in his philosophy, it becomes an inquiry into how beings come to be via *poiesis* (particularly in terms of human *poiesis*, and so, as language or more generally, as inscription). His critique of metaphysics in *Being and Time* is a critique of what we have called a "strong" documentarity, as it is practiced in a phenomenological reduction to consciousness: the use of essentialist and universal categories for determining the nature of beings. Heidegger's later works subject the seeking of a fundamental ontology to an even greater critique, that of *poiesis* or inscriptionality. Heidegger's notion of inscriptionality as *poiesis* engages poetry as a textual form, but as we see in his critique of modern technology and modernity, *poiesis* has critical expanse upon broader social and cultural inscriptions.

Following Heidegger to a certain extent, the argument in this book is that inscriptionality, as a form of representational truth, has taken an historical trajectory following metaphysics, from *a priori* through *a posteriori* modes of representing beings as evidence of essential properties. Documentarity, as the metaphysics that guides inscription in the name of truth, has "strong" and "weak" forms claiming to represent the essence of the entity from the aspect of *a priori* categories of judgment (reference) to the entity's own sense as expressed, and then, as we will see, categorically represented "empirical" affects ("sense").

Heidegger's works never got beyond a generalized critique of ontology as metaphysics through a notion of *poiesis* as materially manifest inscriptional form. And his examples of poiesis in art always remained in the realm of representation, never formalism. The modes of expressive inscription were not investigated as such, at least beyond the project of a fundamental ontology in *Being and Time*.

Bruno Latour's works, for many years, however, have put the emphasis upon the social "machinery" of inscription itself as the mechanism of ontology, and representation is seen in his works as but one mode of inscriptional creation and use. Latour's works put the stress upon *poiesis* as a form of *techne* rather than Heidegger's reverse formulation. We will now turn to Latour's works.

Inscription in the Works of Bruno Latour

In Heidegger's works, thinking inscriptionality begins the process of thinking being in a nonphilosophical manner, which for Heidegger means outside of at least a positivist notion of science and in a genre mode closer to what he identifies as poetry. But is it necessary to leave "science" in order to find *poiesis*? For Bruno Latour, it is precisely in the doing of science (and for Latour, science is always a doing) that one finds the best example of ontology as inscriptionality.

As Foucault (1971) argued, naming in natural history in early modernity was meant as a demonstrative procedure upon a singular entity: name and thing were indexed to one another in a mode of representational correspondence, via a universal category of a thing's essence, which had morphological, behavioral, locational, or other traits. Tabular classification is one form of classification; classification, taxonomies, and ontologies are modes of analytical naming and have a type-token relationship that we have called "strong documentarity."

Natural philosophy in the seventeenth and eighteenth centuries broke from the medieval theological view of the world as an intertwined referential totality by introducing a concern with natural entities as distinct from the social and cultural uses of them. Natural entities still signify, but what they signify is indexed to their own powers as morphological types. Natural philosophy saw the being of an entity displayed by gross external appearances; evolution would later see the signs for the being of an entity as embodied in sometimes-small anatomical organs and their functions that were clues to a genealogical inheritance. Correspondingly, the basis for identity shifted from that of differences marked by synchronic tables to functional organs as indicators of genealogical inheritance (Foucault, 1971; Tort, 2001). With the emergence of biology as a science, beings and their species types become singular within branching trees of inheritance and environmental selection of favorable traits, rather than transcendentally individual; chance differences in individual dispositions are released as evolutionary powers for a species' future "development" because of changes in environmental affordances. Psychology in human beings developed in the nineteenth and twentieth centuries as a science of mind, sometimes attributing greater causation to biological, social, and cultural affordances, with "innate" or "inherited" dispositions or "traits" taken to be due to biological or "environmental" (i.e., learning and development) causes.

Bruno Latour's notion of "inscription," in contrast to "sign," is useful in considering social, cultural, and physical affordances and how these come together in institutionally mediated modes of science. Whereas eighteenth-century natural philosophy, and later positivism, took particular entities out of a medieval epistemology of social and cultural networks of symbolic reference and gave them their own powers as entities (though as representatives of universal types or "facts"), Latour's works reassemble the social and cultural elements that let beings appear as empirical through different modes of scientific tools of inscription that are not reducible to "sign" or documentary traces. The social and cultural rejoin physical (and with this, environmental) affordances in our recognition of entities within scientific research.

In this section, I will examine two of Latour's works in order to understand this analytic advance of inscriptions over sign-based reference. In Latour's works, the term "inscription" connotes a mode of marking that joins performance and description (and also prescription), rather than the term "sign" which connotes in his works representation (as we will discuss in terms of Latour's understanding of maps). "Inscription" has a tool-like connotation in his works, whereas "sign" has an ideational connotation. For Latour, scientific evidence is a product of the inscriptional mechanism of research tools and institutions.

The two works that I will concentrate on are Latour's recent book, translated into English as *An Inquiry into Modes of Existence: An Anthropology of the Moderns* (Latour, 2013), and an earlier work that specifically discusses documents, which remains still in French as of this writing, "Ces réseaux que la raison ignore: Laboratoires, bibliothèques, collections" (Latour, 1996).

An Inquiry into Modes of Existence is a grand work in both size and content. In some ways this book revises Latour's 1991 book *We Have Never Been Modern*, attempting, as Latour states in the beginning of the book, to give a more affirmative explanation to the critiques of his earlier works. Uniquely among his works, however, it attempts to merge a constructivist or actor network model with a theory of beings as modes of expression whose truth-values are determined by their different "felicity" conditions. Through the notion of modes, some sense of dispositions is presupposed, though, at least in my reading, the ontology for this is rather murky in that Latour wants to avoid a notion of innate dispositional powers on the one hand (at least as embodied in any notion of substance), while also avoiding a phenomenology of sheer inscription.

Latour's (2013) book also covers a great deal of modalities of expression, from technology to imagined beings and spirits, to literature, law,

The Philosophical Perspective

and politics. This broad range purposefully blurs the distinction between the "two cultures" of science and the humanities (as Latour's works often do), but it leaves us in a difficult place when we try to account for different modalities of expression among beings (not least between imagined, social, and real entities), effectively turning all entities into the products of inscriptions, and so, perhaps, turning inscriptions back into signs. The possible danger is that reducing all *poiesis* to sociocultural technologies and institutions of inscription comes at the cost of losing ontology to epistemology.

An Inquiry into Modes of Existence uses the narrative device of a female ethnologist researching the discursive behavior of "the moderns." Latour argues for examining the specification of modes of beings across various ontological types across the sciences and humanities. The very blurring of these ontological types is purposeful, however, as Latour rejects a materialism whereby objects can be said to exist, just as much as subjects. Instead, both of what Latour's "moderns" call "subjects" and "objects" are said to be the products of networks of knowing.

The following extended quotation gives us a sense of the epistemic relation between actor networks and modes of expression in Latour's book. Rather than relying on a theory of dispositions lying in innate qualities, Latour's theory largely remains within the domains of a process philosophy that melds actor network theory with Whitehead's notion of entities as occasions for transformations of affects. Modes of expression are lines of force in inherited processes:

> Fortunately, the anthropologist of the Moderns is now equipped with a questionnaire that allows her to determine TRAJECTORIES fairly precisely without having to involve them in the major issue of OBJECTS and SUBJECTS (from here on always in capital letters as a reminder that we are steadily distancing ourselves from them). Every instance of continuity is achieved through a discontinuity, a HIATUS; every leap across a discontinuity represents a risk taken that may succeed or fail; there are thus FELICITY and INFELICITY CONDITIONS proper to each mode; the result of this passage, of this more or less successful leap, is a flow, a network, a movement, a wake left behind that will make it possible to define a particular form of existence, and, consequently, particular BEINGS.
>
> When we use this questionnaire with beings of reproduction, we understand why it would be very unsatisfactory to qualify them by saying that they form a simple "material world" or that they are "PRELINGUISTIC." On the contrary, they express themselves, they predicate themselves, they enunciate themselves, they articulate themselves admirably. To be sure, they reproduce themselves

almost identically, but that is not reason to believe that they do not have to pay for maintaining themselves in existence by passing through other beings, thus by a particular PASS. Indeed, this is probably what qualifies them best: they insist on existing *without any possibility of return*. The risk they take in order to continue in existence can never be taken a second time; if they fail, they disappear for good. No mode is more demanding in terms of the difference between success and failure.

We can recognize them first in two forms, as LINES OF FORCE and as LINEAGES, two distinct ways of defining the minuscule or massive hiatus that separates their antecedents from their consequents. This difference between these two types of alignments is well marked by Whitehead when he points out humorously that museums of natural science keep crystals in glass cases, but they have to keep living creatures in zoos and feed them!

The insistence proper to lines of force—these entities called, too disparagingly, "inert beings"—has repetition and quantity as its consequences; they are numerous, no, they are countless, *because* they repeat themselves and insist. The very notion of FORCE, which will be such a useful handhold when physics and then chemistry are born, is the consequence of this repeated insistence and this proliferation. But if these entities form *lines*, alignments, it is because, despite the hiatus, despite the leap from one instant to the next (a leap impossible for human eyes to discern), each occasion inherits something that allows it to sketch out, as Whitehead says (he was their mentor and, as it were, their protector!), "historic routes." The notion of a "material world" would be very ill suited to capturing their originality, their activity, and especially their diffusion, for it would transform into a full, homogenous domain what has to remain a deployment within a network of lines of force. (Latour, 2013, pp. 100–101)

If, by "force," we are to think of the powers of entities to express themselves and by "lineages" their ability to repeat themselves from one Whiteheadian occasion or event to the next (i.e., when real entities are expressed), then does this exhaust the powers that form the expression of entities? In other words, are the expressions of entities limited to the modalities that are specific—Latour's "specifications"—of them?

Latour's response, as one would expect, is no. There are other "material" conditions that affect the powers of singular entities and align them as mutually constituting expressions, which give them what Latour, in his chapter on beings of fiction (chapter 9), calls "sense" (echoing Deleuze's lengthy discussion of literary language in his *The Logic of Sense*, which itself owes a debt to Frege's distinction between sense and reference). *Sense* means not just "affect," but also the directionality of affects that result in subsequent states of entities and events and then *their* subsequent affects and effects. Sense is meaning that has a directionality for expression and agency, not just a meaning as a result of a referent.

Latour wishes to avoid a reduction of entities to being substances in order to avoid transcendental philosophy. For this reason, Latour grounds modality in process philosophy, which, however, makes any analysis of modal types difficult to arrive at outside of inscriptionality (e.g., using the concept of innate dispositions). Consequently, contextual affordances for expression in Latour's work are quite broad in terms of their analytic detail. Lacking a theory of dispositions other than modalities, we are left with a notion of substance based on processes or "reproduction." Latour continues his argument:

> But the grasp of existents according to the mode of reproduction is not limited to lines of force and lineages; it concerns everything that maintains itself: language, bodies, ideas, and of course, great institutions. The price to pay for the discovery of such a hiatus is not as great as it appears, if we are willing to consider the alternative: we would have to posit a substance lying behind or beneath them to explain their substance. We would certainly not gain in intelligibility, since the enigma would simply be pushed one step further: we would have to find out what lies beneath that substance itself and, from one aporia to another, through an infinite regression that is well known in the history of philosophy, we would end up in a Substance alone, in short, the exact opposite of the place we had wanted to reach. It is more economical, more rational, more logical, simpler, more elegant—if less obvious in the early phases owing to our (bad) habits of thought—to say that subsistence always pays for itself in alteration, precisely for want of the possibility of being backed up by a substance. . . . No TRANSCENDENCE but the hiatus of reproduction. (2013, p. 102)

For Latour, ontological categories function not as transcendental substances to contain entities, but rather as "grammatical" elements within use in order to guide sense; what Latour calls in his book "prepositions"—category names—that give us directionality in the use and expectations of materials. Latour's category prepositions are another type of grammatical signpost in using and understanding things, telling us what to expect of a being, rather than what it "is" in a transcendental sense. (E.g., using a grammatical preposition that we will discuss later, "of": this preposition would not operate as a sign of something being possessed by something "larger" or previous that we suppose exists—say, an individual antelope *of* a transcendental "class," or Platonic *eidos*, of antelopes—but rather, as a signpost of "belonging to" a class, whose existence is solely that of a function within the heuristics of a practice called "taxonomy.") Latour has always been a fierce critic of what I have called strong documentarity as ontology, though it is allowable as an epistemology, just so we don't slip into an ontological reification of classes. As often occur in his works, a performative and even prescriptive, rather than a representationally descriptive, understanding of maps and signposts is offered as an example of Latour's critique. For example, in the below quote showing the role of what he calls "prepositions" in the function of a map of Mount Aiguille:

> Appearance allows itself to be seen in the *direction* given by the preposition, like the path followed by a hiker who is reassured but nevertheless careful not to make a mistake. To follow this direction really amounts to leaving the placard behind, heading in the direction it has indicated, without there being in this forgetting the slightest denial of the direction it has indeed *given* you. No one will say that the term "novel," "provisional report," or "documentary fiction" on the first page (appropriately called in French the *page de garde*, the "warning" page) "founds" the reality of the volume that follows, but no one would think of saying that a signpost obscures, contradicts, denies the direction it designates, no one can claim, either, that it would be much more rational to do without any signs at all. In other words, we must seek neither to get rid of appearances nor to "save appearances"—to save face—nor to traverse appearances. We must simply head in the direction indicated by the preposition, without forgetting it. Appearances are not shams. They are simply true or false depending on whether they veil or lose what has launched them. (2013, p. 271)

The reference to the hiker in the above quote echoes an important example in chapter 3 of his book, which concerns Latour's notion of reference as being made up of practical links or "chains of reference": a map of Mont Aiguille, Latour argues, has its referential function not by virtue of a descriptive correspondence of map (literal or mental) and thing (as an instance of *adaequatio rei et intellectus*), but rather by a pragmatic alignment of the coordinates and symbols on a map with the geography of a mountain for the purpose of successfully hiking the mountain. Latour's work is grounded in a pragmatic understanding of signs, based in use, much like Wittgenstein's philosophy of language. Evidence is the result of processes and institutions that inscribe what is viewed as the most productive way of moving forward—in hiking or in doing research. Latour is very much a "pragmatist."

What are included in the "infrastructure" of a map—this "immutable mobile," to use Latour's famous term—are the very cultural references, the

The Philosophical Perspective

very human labor, the very social and economic productions, that make such a device useful and habitual. Inscriptions guide minds and their relations to the reality of the world. Latour writes:

> In the first sense, the expression "immutable mobiles" sums up the efforts of the history and sociology of the sciences to document the development of the technologies of visualization and INSCRIPTION that are at the heart of scientific life, from the timid origins of Greek geometry—without trigonometry, no topographical maps—up to its impressive extension today (think GPS); in the second sense, the same expression designates the *final result* of a correspondence that takes place *without* any discernible discontinuity. Quite clearly, the two meanings are *both true at the same time*, since the effects of the discontinuous series of markers has as its final product the continuous itinerary of the sightings that makes it possible to reach remote beings without a hitch—but only when everything is in place. This is what I said earlier about the two meanings of the word "network": once everything is working without a hitch, we can say about correspondence what we would say about natural gas, or WiFi: "Reference on every floor." (2013, p. 77)

Latour's notion of "inscription" here is very important, both for Latour's works and for the trajectories of the indexicality of ontological representation that we are following in this book. With this term, Latour packs together both signs and the social, cultural, and physical affordances and processes that give meaning to those signs within a temporal horizon.

Latour's claim is that his pragmatic vision of science erases the divisions of society and nature. We can no longer have truth only in nature, only in society and culture, or only in "language." Instead, entities come to appearance through the affordances of habits, expectations, and methods of knowledge inscriptions. Such a "constructivist" claim can be taken quite radically, so, for example, that the natural world is deprived of any ontological being outside of human understanding (see Neyrat, 2018, for his criticism of Latour on this point).

Libraries and Other Data Centers

Earlier than *An Inquiry into Modes of Existence*, in his article "Ces réseaux que la raison ignore: Laboratoires, bibliothèques, collections" (1996), Latour discusses inscriptions in chains of reference, from the perspective of maps, laboratories, natural history collections, and, importantly for our purposes, libraries. Through his famous notion of a "center of calculation" (*centre *de calcul*) (or less literally and more colloquially translated into English as "data center" or "computer center"), libraries, within a bibliophilic tradition, are displaced as the privileged centers for knowledge. Libraries and data centers (or rather, as Latour sees them, libraries *as* data centers) are just one among many nodes in networks of knowledge production. In his article, Latour writes:

> I will not follow the path which leads from one text to another in the interior of a library, but the path which leads from the world to inscription, up and downstream from that which I would call the "center of calculation." Instead of considering the library as an isolated fortress or as a paper tiger, I will depict it as the node of a vast network where there circulate not just signs and not just matter, but matter becoming signs. The library doesn't rise as a Palace of Winds, isolated in a real landscape—too real—which serves as a setting. It curves space and time around it and it acts as a temporary receptacle of *dispatching*, transforming, and switching the very concrete flows it continually creates. (1996, p. 24; my translation)

A library in this sense collects information, transforms it, and then dispatches it again. It is a data center, no more, no less. It is not the final, textual, resting place for processes of knowledge production.

What, for Latour, is information and how does it relate to inscription? Latour writes,

> Let's start with the sign and how to define information. Information is not a sign, but a *relation* established between two places, the first which become a periphery and the second, a *center*, a condition that between the two circulate a *vehicle* that one often calls a form, but that, insisting on its material aspect, I will call an inscription. (1996, p. 25; my translation)

"Information," for Latour, means "inscription," and inscription is not necessarily descriptive representation, but more generally, it is the inscription of a relation between sociocultural institutions and the entities that such study. Such inscriptions require working across institutional, material, and formal disjunctions. While some elements of the "periphery" are reduced in a documentary collection, other elements are amplified or "capitalized." Institutional documentary collections are, in a sense, constituted by the making of models and "smoothing"[5] them into an average of acceptable knowledge for further use "downstream."[6] An example of such is individual wild birds being captured and brought together as types within an aviary or taxonomic collection. But books and libraries are examples, as well. Such examples are then used for research consultation, further taxonomic development, education, and so on.

As Latour writes:

> But, compared to the original situation, where each bird invisibly flew in the confusion of a tropical night or a small polar day, what fantastic gain, what an increase [i.e., when the birds are collected in an aviary]! The ornithologist can then tranquilly and comfortably compare the pertinent traits of thousands of birds through the immobility, by the exposition, by the naturalization of the exhibited birds. What lives dispersed in singular states in the world are unified, universalized, under the precise gaze of the naturalist. (1996, p. 27)

Inscriptions, according to Latour, "take responsibility for matter"; indeed, they are given responsibility by matter (*se chargent de matière*) for their appearance as knowledge. When reading a map, all of the social and political tools and activities—standardized signs, journeys to other lands, learning how to read maps—come into play, along with the material object "itself" (Latour, 1996, p. 43). Latour's point is that the creation of meaning in documents requires many different social, cultural, and material networks or assemblages put together for the purpose of creating such appearances. The entity comes into appearance as a sort of focal or indexical point within assemblages of knowledge-creation tools and institutions. Such, what we can call, after Heidegger, *poiesis*, *is* information. These materials and this labor together make up the entity's "information."

Latour's inscription of libraries, too, as centers of calculation, is very important for our understanding not only of libraries, but also of the types of documents that both libraries and nonlibraries collect. In this understanding, libraries and other documentation centers are not seen as simply endpoints of information, knowledge, or what is true. Rather, their materials are understood as parts of material processes of knowledge creation and distribution.

In sum, Latour's skepticism regarding a philosophy of substance stems from his critique of representation as being the end goal of scientific processes. As we will see when discussing Paul Otlet's theory of documentation, documentation and libraries are sometimes seen as containing representations of facts of the world; they supposedly represent substances and the substantiality of the world as a totality (e.g., in the universality of some modern "knowledge organization" classification systems, such as the Library of Congress or the Dewey classification systems). The downside of

Latour's skepticism is that of potentially reducing all ontologies to being products of human inscription. More recently, however, Latour's (2013) turn to modality attempts to reassert ontology, though again through process philosophy inscriptionality and without a return to what he sees as the philosophy of substance.

What is needed in Latour's work, I would suggest, is a more fine-grained notion of modalities for expression, particularly one that better accounts for powerful particulars outside of human inscriptionality. Latour's (2013) work heads in this direction, but I don't think it quite accomplishes it partly due to his epistemological groundings in process ontologies, such as actor network theory and Whitehead's philosophy, and partly due to his fear of eliciting a philosophy of substance. Ontological substances need not be transcendental or even universal, however, and they need not end in a philosophy of documentary signs, either. We can have a theory of substances made up of assemblages of innate dispositional powers, with "innateness" varying from relatively fixed properties of powers with inorganic entities to learned ones with organic entities. With any of these, it is beyond dispute that human modes of inscription are necessary for human knowledge of such entities to occur, but such a view doesn't obscure the fact that these powers also reside in the entities independently. Indeed, that's partly why we have and need technical and technological interventions into observing such powers, why we need science. We can have a realism that is not divorced from institutional pragmatism, and I would argue that we must have such in order to make sense of science as something other than systems—or even simply "habits" or practices—of signs.

Rom Harré's Theory of Expressive Powers

For what I see as a more nuanced view of ontological modalities, I would assert the usefulness of dispositional theories toward demarcating powers of expression in both groups and individual powerful particulars. Dispositional theories make use of many different types of terminology, but for our purposes I will make use of dispositions understood as expressive "powers," following Rom Harré's (particularly Harré, 1995) many works in both the physical and social sciences. Following Harré, I will use an expanded notion of James J. Gibson's notion of affordances to discuss what Harré calls "powerful particulars" and their modalities of expressing being. The warrant for this expansion of Gibson's notion is that of using a nuanced dispositional-affordance theory of expression to analyze beings and their evidentiary expressions across social, cultural, and physical regimes. Dispositional theories of powers for expression are useful in order to address beings as both entities and modes of entities, and so with this epistemic toolbox we can begin addressing beings as something other than signs or even inscriptions of social and cultural systems, namely, as expressive powerful particulars.

First, though, what do we mean by "power"?

In the post-Foucaultian world of critical discourse, the term "power" in English often elicits notions of repression. However, the word "power" in English has two distinct senses (reflected in the Romance languages by two different words; for example, *puissance* and *pouvoir* in French, or *potenza* and *potere* in Italian): "power," in the sense of a generative or what we'll call an "expressive" power that is potential or actualized (like a plane at takeoff), and "power" in the sense of a repressive power. When we speak of what Harré has called "powerful particulars," we are speaking of power more in the sense of the former. Such powers can, of course, in some conditions, act repressively upon others (and upon other powers of the entity expressing them), but the two senses of the word are distinct.

Powers in this first sense allow expressions. But this has to be understood in two ways: the dispositional powers (*dispositions*) of an entity to express itself and the "contextual" *affordances* that allow it to do so. With an entity of inert matter, the innate physical properties of a thing can be seen as relatively fixed (at least at certain physical levels—e.g., non-quantum), and so one of its modalities is that of very stable expressions within any given set of affordances. With the right conditions, volatile chemicals *must* ignite, otherwise they are not pure or the conditions are not right. Organic beings, composed of multiple, developing, parts can adapt to changing external affordances and so they *may* act this way or that way. An organic entity can acquire and learn to express this or that set of powers in the face of different sets of affordances; that is, it acquires new dispositions.

Higher-level organisms obtain large and experientially derived unique "tool boxes" of dispositional skills (and so, of potential expressions) for various situations. After Harré, we could call the unique toolboxes of humans (or even of any higher-level organic being) a "self," and distinguish this, at least in the case of human beings, from his notion of "person" (which would indicate the social rule-based expectations for behaviors upon an individual and the individual enacting such; Harré, 1989). From this perspective, selves choose their expressions from repertoires that are in common, but individually learned and possessed. Each individual has a unique set of tools for expressions that constitute a self.

How dispositional powers belong to entities depends not only on the type or "mode" that it is, but, particularly for higher-order organic entities, the uniquely particular entity that it is or has learned to be. In terms of evolutionary theory, each individual is essentially singular, and types are identified by common parts, functions, and expressions, which, of course, depend on the typology or genealogy used to judge the entities.

As has been earlier suggested when discussing Heidegger's notion of *aition*, there are three types of "material" affordances that we may consider for an entity's expressions: physical, cultural, and social. "Physical" refers to *bodies* and their forces, "cultural" refers to symbolic *forms* for expression, and "social" refers to social manners or *norms* of deploying cultural expressions. For example, my mouth, larynx, lungs, and tongue form physical elements of my verbal expressions, the English language, the cultural forms for them, and norms of expression in a social situation (e.g., shouting, speaking quietly, using a word like "fire" in an appropriate context) constitutes the social affordances for them.

Social, cultural, and physical affordances are all "material," in the sense that they have coherence, force, and resistance (Harré, 2002). Planets have a necessary physical materiality, whether or not they also have cultural or social meanings. Sulfur may have cultural and social affordances, but these do not belong to the entity itself but to the conditions for its human symbolism and the social means of deploying this symbolism. The same is true for planets. Ideas or concepts, however, as assemblages of words and actions, have social and cultural materiality whatever their distinct physical form (e.g., spoken, written).

The rise of modern empirical natural sciences was characterized by the attempt to understand the physical affordances or "causes" of natural entities outside of their cultural and social meaning. This isn't to deny that such research hasn't been driven by cultural, social, financial, historical, and other reasons, nor is it to deny that what is identified as the physical causes of an entity are not selected by these other affordances. Rather, modern science merely recognizes that the entities themselves, as physical bodies, have innate dispositions.

In terms of the social and cultural sciences, there has been a tendency, however, in modern science toward a reduction of all powers to being seen as causally reducible to physical dispositions, but this physical reductionism is misdirected when analyzing the social and cultural qualities of higher-level organisms. Social and cultural materials compose the "minds" of individuals. Social and cultural affordances can only be *correlated* with physical affordances, not causally reduced to them. Brains are composed of neurons and other physiological entities; minds and their expressions and ideas exist through culturally and socially made expressions.

Human bodies can act as physical means for making social and cultural expressions, and these latter affordances may create habits that then reciprocally reshape the physical dispositions and so the affordances of a body. So, for example, baseball pitchers' ability to throw a curveball cannot be found solely in their arm muscles or any other part of their body. Rather, it also belongs to the very notion of there being a cultural form called a "curveball" in a social function called "playing baseball," and these affordances together can then modify a person's body into having the physical capabilities of a baseball pitcher.

Both throwing a baseball and doing philosophy, for example, are skills, and both involve overlapping social, cultural, and physical affordances. How we put the emphasis upon whether one activity is an "intellectual" or is a "physical" activity may depend on what we see as the primary and necessary skills for a given expression are and so the likely causal site of important dispositions. In some cases, judgments of this may change depending on different situations and different epistemologies. Where medieval or Elizabethan scholarly observation saw wild natural entities as greatly dependent on moral attributes for their characteristic expressive powers of being, the modern natural sciences may see such entities as being less so. For example, where at one time the salmon's migration upstream showed sacrifice and nobility, today we look at such a perspective as being an anthropomorphic reading of salmon breeding behavior.

The powers of empirical entities are recognizable as belonging to empirical entities because they resist and sometimes defy us in understanding them. Natural entities, particularly inorganic ones, take on a fixed character of expression no matter what we call them (e.g., if we call this entity "sulfur," or if we call it by the chemical element symbol of "S," it still has the expressive properties of that substance). We say here that the entity has a strong *reference*, and a weaker (social and cultural) *sense*, because its expression is less influenced by the term we use of it and more influenced by its own innate qualities of expression. On the other hand, higher-level organisms may have a strong sense of expression, too, according to their own cultural forms and social norms (with humans, certainly, but also in the case of other higher-level organisms). My domestic cats may decide to come to my call or not, seemingly dependent on their individual moods and life histories, whether I, as their owner, prefer this or not.

At least with higher organisms, what a substance is—according to what expressions we witness from it—must always have some degree of provisions that what we witness may not be definitive. Since these are sociocultural "objects" (or more properly, subjects), how we describe a person, for example, may vary quite a lot across observers, the language choices they have to describe another person or event, and the activities of the subject under observation at a certain time. All human beings appear as geniuses and morons, both because other people call them such and because they act as such in different situations.

Some substances can never be observed and are only hypothetical presuppositions of what we suppose are their expressive powers. Sometimes these presuppositions refer to real entities (e.g., black holes); other times they refer to nominal entities (e.g., social science entities, such as "minds") or even to ethical principles or regulative ideals (e.g., in Kant's practical philosophy, his categorical imperative).

A theory of dispositional powers and their affordances allows us to gain a level of specificity and stability in regard to our understanding of the expressions of individuals and groups. It allows us to concentrate on particulars in such a way that they are recognized as powers that are not fully dependent on either *a priori* categories of judgment or on sociocultural inscriptions for their powers. Evidence can be understood as reflective of the powers of a particular entity *along with* their interpretation and inscription.

The trade-off in regard to strong documentarity is that we must forgo absolute certainty in regard to essence, and instead appeal to expressive powers and our understanding of them, and so, to probability. Substance, here, is not transcendental; rather, it is dispositional. The trade-off in regard to inscription is that we must accept substance as existing to various degrees independent of human activities. Substances with inorganic entities at normal levels of observation tend to have strong correlations between dispositions and expressions, so strong that we say that there is a causal relation. In the case of particularly higher-level organic entities, the relationship between hypothesized dispositions and expressions is causally weaker, and must be asserted through correlations, preferably with observations over time.

We make loosely "correlational" assertions about people's social and cultural attributes all the time in the case of moral judgments, where we are often describing people's moral substance. "Maturity," however, is the developmental process of realizing that attributes of selfhood apply to other persons, as well as that personhood applies to one's self. And since selves are defined by potential expressions and persons defined by socially expected or "necessary" expressions, hypotheses and correlations apply more to the latter than to the former, which can, in an analytical manner, be seen as being more causal.[7]

In sum, an ontology of substance based on dispositions and affordances can give us tools for better understanding the power of empirical entities across the sciences and how they become evident to us. In the following chapters, we will see different forms or genres of and modalities for evidential and self-evidential expressions across a spectrum of *a priori* and *a posteriori* methods. In earlier chapters, we will see evidence appear more through semiotics and in later chapters more through empirical powers, though we will end the book by looking at these empirical powers being recaptured in communicative and predictive computational systems. Throughout the book we will be looking at inscriptional genres as information producing technologies, that is, technologies of entranceways and exits for expression, and thus, technologies of judgment for what is, what was, and what is to be.

2 Documentarity in the Works of Paul Otlet and Georges Bataille: Two Competing Notions of "Document" and Evidence

> If one wants to see civilization, one would be better off visiting a slaughterhouse than a museum.

The ubiquity of strong documentarity in Western cultural traditions leads to the question of whether there has been a countertradition to such, less based on particulars evidencing essential and universal type-classes and more based on self-evidence. It also leads to the question of the relationship of documentation or library science theory to philosophical modes of documentarity.

At the end of the nineteenth century and the beginning of the twentieth century documentarity took two extreme epistemic forms: that of positivist representation (stressing reference) and that of a counter "materialism" (stressing sense) of entities. Nietzschean and similar projects of a "reevaluation of all [Western or European] values" (i.e., the critical unmasking of the Western cultural tradition as a "will to power") had far-reaching implications in philosophy, art, and the human sciences during the early part of the twentieth century. Such critiques, together with the cultural ruptures that the First World War wrought to the ideals of "European" elite culture, starkly brought forth two poles for documentarity tied to two forms of knowledge: knowledge as collections of facts about the world, and knowledge as experience, where terms such as "sense," "sensuality," and "the body" played important roles.

Nineteenth- and early twentieth-century library and museum documentation theory, as an epistemology of knowledge that held that the world was represented in collections, was challenged in the early twentieth century by field anthropological research and ethnology. Science, understood as a distantly held documentary collection of knowledge, was challenged by science as a practice of investigation and inscription, while at the same time, the latter often making the same type of representational claims of realism in documentary field research products (such as notes, still photographs, and film recordings). Generally speaking, evidence was relocated from what was literally or figuratively seen as storehouses of facts (books, libraries) to factual experience (empirical research, "the body"). Again, however, experiential inscriptions often took on the same metaphysical claims for representing the real. Here we have a case of metaphysics and representation coloring not only strong, but also weak, "empirically" based, documentarity. As we will see throughout this book, the same (though with less brash a tone) dialectics, perhaps most pronounced when explicitly involving documentation (or more generally, information) technologies, occur throughout the twentieth and into the twenty-first centuries.

We can see this by contrasting the works of two contemporary librarians/documentalists writing in French in the early twentieth century: Paul Otlet, who is seen as the father of European documentation, and the writer and philosopher Georges Bataille, whose work was heavily influenced by early and developing French field anthropology, ethnography, and ethnology. By contrasting Otlet's utopian positivist understanding of library documentation with Bataille's dystopian "base materialism," we can see two seemingly opposed epistemologies—one grounded in ideal reference and the other in materialist sense—each engaging in documentary (and, in Otlet's case, at least, institutional) practices of representation. In examining Bataille's practice and reading of documents as valorizations of "the body," we may recall Heidegger's critique of Nietzsche's reevaluation of all values: that by replacing mind with body, reference with sense, and being with becoming, Nietzsche (and we may add, Bataille after him) had merely inverted Platonism, rather than having destroyed it (Heidegger, 1979).

We have case of this, too, in early and mid-twentieth-century French ethnology, the earlier of which had a strong influence on Bataille's works. As Vincent Debaene has shown in his book *Far Afield: French Anthropology between Science and Literature* (Debaene, 2014), the collections orientation for anthropology developed in the nineteenth century with the practices of museum collections.[1] Museums were seen as collections of cultural knowledge, much like libraries were and largely still are: distant from their points of construction and arranged more from the perspective of their collectors than by the expressive powers of their objects or subjects of study. They were, in a strong documentary and colonial sense, European "centers of calculation" for collecting and measuring others in distant lands. Like their offspring, ethnological exhibitions or "human zoos" in the late nineteenth and early twentieth centuries (live exhibits of other cultures by representatives of those cultures), museums had mixed epistemic, education, and entertainment value in a fascination with the "other."

As Debaene (2014) recounts, by the 1920s, however, field anthropology and early ethnology developed in France as a critique against this documentation orientation in museums. Like their earlier British and American brethren, anthropologists took to the field, and by the 1950s, according to Debaene, French anthropologists were writing two books, one of a scientific representation of the studied culture and the other of their experiences during that research. The first book depicted the studied culture in a structural method similar to museum classifications: as the "objective" cultural differences of one cultural group as compared to another, within systems of identity and difference. The second book, however, was more of an experiential narrative of being "in the field and in the flesh" as it were. Nevertheless, as Debaene notes, both of these works—a type of strong documentarity and a type of weak documentarity—were representational.[2] They both represented the "other": the first in a mode of scientific objectivity, and the second in a more phenomenological and less "methodological" frame.

Debaene writes:

> We have seen that, in the 1930s and beyond, the anthropologist's "second book" in most cases constituted an attempt to restore the "atmosphere" of the society under study. This brings two elements to the fore: first, naturally, the inadequacy of the documentary and museological paradigm as it was applied to social realities; second, the fundamental unity of the project underpinning both of the ethnographer's two books, for whether we consider the first or second of these texts, at stake is always the description and restoration of a "real" cultural object. Put another way, the project of anthropology as it was formulated in the interwar period remained fundamentally empiricist: it held that the ethnographer's objects are entire societies and that these had to be fully described. Regardless of whether this object was material or mental or whether one hesitated between different methodologies (collection or immersion) and different modes of writing up the results (a scientific study or a "literary" work), it hardly mattered in the end since the same epistemology governed both the scholarly presentation of scientific data and the "evocative" ethnographic narrative. (2014, p. 276)

In this book, I argue that the metaphysics of documentarity not only possess an idealist paradigm, but also historically evolve toward an attempt to represent the particular as particular, however paradoxical this might be, in terms of experiential or "empirical" methods of sense. In this, the aesthetic or experiential index of documentary (broadly speaking) representation shifts, from a mimetic to a metonymic relationship between the universal and the particular (as we will examine in the next chapter), and finally evolves to a synthesis of tracking the particular in terms of expected inductive paradigms, as well as innate powers and habits. In Otlet's works, we see classic "strong" documentarity in a positivist paradigm led by library-held "facts." In Bataille's works, we see a series of tropic and literary inversions of the Otletian paradigm for documents, while the "other" remains an object of erotic fascination, inclusive, if not to say dominant, within the author's own self and culture.

Paul Otlet's Bibliographic Positivism

Paul Otlet was a rather forgotten figure after his death in 1944 until his work was recovered by his biographer, W. Boyd Rayward, some forty years later. Today, Otlet is widely celebrated as an information science predecessor and a visionary of something like today's World Wide Web. His vision of science as documentary knowledge representation continues today, however, not only in libraries, but also in information technology firms claiming to be "keepers of knowledge" and in representationalist claims for information visualization (performed in Börner, 2015, and critically discussed in Drucker, 2014).

Paul Otlet is sometimes referred to as the father of the European documentation movement, a predecessor to American documentation and its successor of information science. He was coinventor with Henri La Fontaine of the Universal Decimal Classification scheme in 1905. He was a prolific and voluminous author who wrote on the social need for international institutions—foremost documentary intuitions, but also governmental, monetary, and educational institutions—and he was the founder of a large world library and museum in Brussels between the First and Second World Wars, the Mundaneum.

As Balnaves and Willson (2011) argue, what they call the "Otlet" tradition of information differs from what they call the "Cutter" tradition, because in the latter, it is the material item (e.g., the paper document or the virtual document) that is seen as being "information," whereas in the Otlet tradition, it is the *content* of such that is seen as being "information" (a very important emphasis when it comes to digital content). This tradition of seeing the "content" of a document, regardless of its material form, as being what constitutes its "information" continues up through contemporary information science.

For Otlet, documents, in the sense of the above notion of information content, are representations of the world. What documents contain is information, and the information of documents is "authoritative" in a strong sense—documents contain or represent (depending on one's grammar) "the facts" of the world.

Otlet's theory of documentation follows a picture theory of language, similar to then near-contemporary works in positivist philosophy such as Ludwig Wittgenstein's *Tractatus Logico-Philosophicus*, with documents and knowledge-organization systems (such as classification systems) playing the role of atomic statements. (Though it should be noted that, according to Otlet's "monographic principle," larger documents may be amendable to being broken up [or in paper form, cut up] into smaller informational documents, where each essential, or "micro," document is a representation of a single "monographic" fact.) Since for Otlet, science is an epistemological domain of true statements of facts about the world (i.e., "information"), documents and library collections hold such "information" in their documents and in their bibliographic organization and the metadata codes for texts (classification codes, cataloging codes, etc.). For Otlet, true knowledge resides in the informational content of documents, and the further "purification" and organization of such occurs in the abstracts and other metadata or "knowledge organization" systems of libraries.

What we could call Otlet's bibliographic atomic positivism can be conveniently seen in the following two illustrations from Otlet's book *Traité de documentation: Le livre sur le livre: Théorie et pratique* (1934). It should be remembered that far from exemplifying a forgotten historical-epistemological moment, Otlet's epistemological commitments continue today in the notion that what information tells us is what is true—or at least is potentially true. (As I will suggest throughout this book, literate societies tend to exhibit a slippery, and sometimes dangerous, slope between literary or graphic taste and epistemic belief.) In Otlet's theoretical works, the materiality of documents and the material, cultural, and social processes that mediate their appearance as information are largely viewed as being inconsequential to the information or knowledge said to be within them, other than when they strengthen the representational claims of such (e.g., in metadata that states the "aboutness" of documents).

In the first illustration, we see knowledge depicted as representations of the world in individual minds and then in various documentary materials, such as books, and most iconically, in photographs. In the second illustration, we see how knowledge (at its "highest" and most "scientific" levels for Otlet) is understood as representations of the world in recorded documents at higher and higher levels of representational abstraction, which for Otlet meant greater essential truths. For Otlet, documentation—meaning

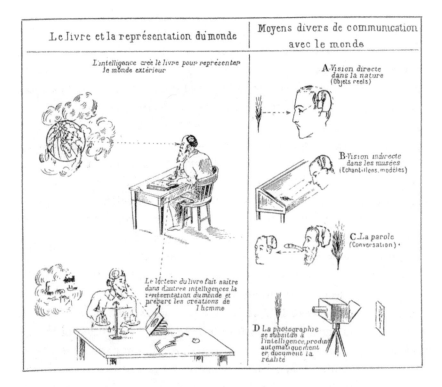

Figure 2.1
On the left, "The book and the representation of the world," and on the right, "Diverse means of communication with the world." From Otlet (1934).

documents and their documentary institutions—represents the true nature of the world. "The book" is Otlet's metonymical trope for all documentation, in part and in totality. And his *Traité de documentation: Le livre sur le livre: théorie et pratique*, as the title tells us, is the book on "the book," and so represents—as theory should, for Otlet—still a higher level of representation, a higher level of synthesis for knowledge, namely, the knowledge of bringing knowledge into being. The highest truths are contained in the most atomic and abstract representations or "metadata" held by libraries: abstracts, ontologies, and taxonomies, and ultimately bibliographic codes (such as classification numbers). Theoretical work on this is something like God's view of all of knowledge (understood as the facts of the world): it includes the "manual" of how it all came and comes about.

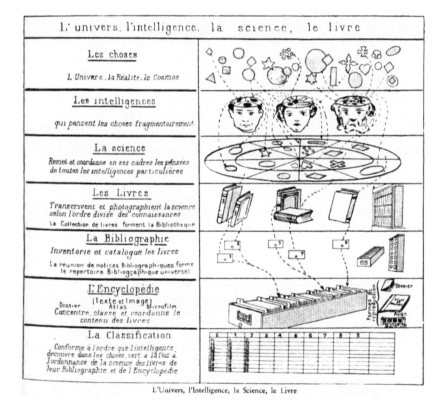

Figure 2.2
"Universe, Intelligence, Science, Book." From Otlet (1934).

Finally, as graphic evidence of what I claimed earlier, namely that Otlet's bibliographic positivism resembles Ludwig Wittgenstein's logical positivism in his *Tractatus Logico-Philosophicus* (Wittgenstein, 1921), let us try substituting "document" and cognate terms for "fact" in the second and third propositions that begin Wittgenstein's book, as I do in the brackets below, to see how well these two philosophies fit together:

1. The world is all that is the case.
1.1. The world [of knowledge, i.e., collections of statements—books, the documentary collection/library/catalog; "*Le Livre*"] is the totality of facts [statements/documents], not of things.
1.11 The world [of knowledge, i.e., collections of statements—books, the documentary collection/library/catalog; "*Le Livre*"] is determined by the facts [statements/documents], and by their being all the facts [statements/documents]).

(adaptation of Wittgenstein, 1921)

Otlet's *Traité* not only repeats the epistemological assumptions of logical positivism expressed in works such as Wittgenstein's *Tractatus*, but it also uses some of the same formal, rhetorical devices Wittgenstein used in his *Tractatus*, namely "atomic" rhetorical units (such as sentence-level statements), which are then built up into more complex rhetorical units (such as paragraphs, book sections, chapters). In this way, and through the use of simple sentences and other rhetorical "monographic" units, both Otlet and Wittgenstein's textual sections build into larger wholes; their texts rhetorically *perform* the analytic-synthetic science that they assert as the true form for knowledge.

In this vein, we can also compare, for example, the similar numeric textual designs that Wittgenstein used in his *Tractatus* (as quoted and amended for content, above) and those used by Otlet in his *Traité*, as shown by the table of contents of the latter (figure 2.3).

In the design of both works, the positivistic claims are not simply made, but rhetorically *shown*. We may recall here Wittgenstein's belief that ethics is most truly shown, not simply said. If this is the case, then both writers demonstrate not only in their epistemic statements, but also in their graphic textual designs, their ethical stances. Similarly, the practice of material culture in library institutions and their theoretical claims are not simply contingent for Otlet, but rather, theory is read out of practices and then comes to reinvest practice with theory in terms of pedagogy, methods, and material and human organization. At least for the notion of professionalism to

.43 Publications de textes 168
.431 Notion 168
.432 Règles pour la publication des textes. 168
.433 Types de publication de recueils ... 169
.44 Commentaires des textes 170

241.5 Catalogues 170

241.6 Tables et tableaux 171

241.7 Autres espèces de documents 174

241.8 Modalités d'une même œuvre. Edition. Traduction. Extraits. Arrangements. 177

.81 Edition 177
.82 Exemplaires 178
.83 Traductions 178
.84 Extraits. Anthologie 181
.85 Arrangement. Transcription 181
.86 Le Neuf et le Plagiat. Emprunt. Copie. Citation 182
.87 Œuvres complètes 183
.88 Continuité des œuvres 183

242 Documents graphiques autres que les ouvrages imprimés. 183

242.1 Les manuscrits 183

.11 Notion 183
.12 Historique 183
.13 Enluminure. Miniature. Décoration ... 184
.14 Erreurs dans la copie 184
.15 Collections. Bibliothèques 184
.16 Catalogue de manuscrits 184
.17 Travaux sur les manuscrits 185
.18 Les papyrus 185
.19 Manuscrits modernes, Incunables 187

242.2 Cartes et plans. Atlas 187

.21 Notions 187

.36 Cartes postales illustrées; Cartes à jouer; Ex-libris 198
.361 Cartes postales illustrées 198
.362 Cartes à jouer 198
.363 Ex-libris 199
.37 Photographie 199
.38 La projection 204
.52 Histoire 210

242.4 Archives, (pièces, collections, dépôts). 207

242.5 Musique 208

.51 Notion 208
.53 Questions fondamentales 210
.54 Instruments de musique 211
.55 Notation musicale 212
.56 Partitions musicales. Bibliographie ... 214
.57 Diffusion de la musique 214
.58 Organisation commerciale de l'édition de la musique 215
.59 Bibliothèque et collection de musique. 215

242.6 Monuments dits figurés : inscriptions, monnaies, médailles 215

.61 Inscriptions 215
.62 Monnaies et médailles 215
.63 Sceaux. Cachets 216

243 Documents dits « Substituts du livre » 216

243 1 Objets. Matériel de démonstration 217

243.2 Disque : phonogramme 220

243.3 Films : cinéma 223

.31 Notion 223
.32 Historique 223
.33 Caractéristiques 223
.331 En général 223
.332 Caractéristiques particulières 224

Figure 2.3

Numerical organization of Otlet's *Traité de documentation: Le livre sur le livre: Théorie et pratique*, from its table of contents. From Otlet (1934).

succeed in modernity, it needs a practical philosophy, for practice without theory is blind, and theory without practice is empty. Both Otlet and Briet were dedicated to this principle in regard to documentary institutions and their relation to knowledge, just as much as the early Wittgenstein sought a role for philosophers to "clean up" the practices of language so that they become as "meaningful" as possible.[3]

Georges Bataille and the Philosophy of Base Materialism

In seeming contrast to Otlet's positivist understanding of documents, we now turn to the work of the French novelist and nonprofessional philosopher Georges Bataille, who worked at the French Bibliothèque Nationale, first as the head of the Department of Medallions and later moved to the Printed Books Department. Bataille was at the Bibliothèque Nationale for nearly twenty years and later worked as a librarian in Orléans and elsewhere.

Though Bataille's literary works are sometimes seen as belonging to the work of the modern avant-garde in literature, it is not because of their formal innovation (there isn't much), but rather because of their pornographic and "extreme" contents. They are representational fiction whose surrealism is in large part due to their extreme sexual content—the surreal lengths that an "empirical" or realist narrative is drawn out into sexual extravagance and how this philosophy of excess constitutes the transcendental or sublime grounds for the real. Indeed, it is only because Otlet's and Bataille's works share claims to representing transcendental essences in documents (one in the form of substances, the other in the form of psychobiological drives) that one couldn't imagine a greater moral and aesthetic contrast than that between Otlet's theoretical works and Bataille's novels and theoretical works as documents of the human world. In Otlet's works, all the writing and all the content it refers to are meant to have clear and distinct boundaries in facts that are rationally organized. In Bataille's novels, excess, disgust, and confusion ooze from every being and their bodily orifices, and all social relations in his theoretical works are somewhat rhetorically meandering. In Nietzschean terms, in their works, Otlet tries to be Apollonian and Bataille tries to be Dionysian, with both, to some degree, becoming the opposite. (Otlet can't stop writing about how simple things are, and Bataille's novels, at least, of course, are filled with sexual fixation to the point of absurdity.)

In Bataille's works, sexual acts, slaughterhouses, war, and other such situations display civilization as based on excess, irrationality, and violence. Whereas documents give evidence of atomic facts and rationality in Otlet's works, in Bataille's works they give evidence of the unreasonable and violent tendencies that led—through subjugation and violence upon those within and outside—to the dominant states of Western Europe and the idea of European civilization. In short, whereas for Otlet documents give evidence of *logos*, documents for Bataille give evidence of the violence that culminates in *logos*.

Further, it is important to note once again not only the rhetorical but also the visual forms that perform the nature of documentary evidence for each writer. For Otlet, the forms of the ideal world were illustrations, perhaps in part because no physical evidence could exist of the epistemic order he saw behind all things and knowledge. In Bataille's works, however, as in field anthropology, photographs offered examples of excess via their contents. So, for example, Bataille used a photograph by Eli Lotar of severed cow hooves neatly lined up outside of a Parisian slaughterhouse for his "Critical Dictionary" section entitled "Abattoir" (Ades, 2006) in the journal *Documents*. For Bataille, if one wants to see the taste of European civilization, one would be better off visiting a slaughterhouse than a museum.

Documents (1929–1930)

Bataille was editor of the Parisian journal *Documents*, which ran from 1929 to 1930. Other than Debaene's (2014) broader work, I am unaware of any critical consideration of the importance of the title of this journal in debates about the social sciences and literature in the early twentieth century. But in the context of Bataille's profession and his friendships in the ethnographic community (such as with Michel Leiris), as well as the subtitle of the journal (*Documents: Doctrines, archéologie, beaux-arts, ethnographie*), it would be hard to think that "document" wasn't a contested term during this time, since what is offered as documents in this journal is certainly not like Otlet's illustrations of rational harmony for the world. Increasingly, under Bataille's editorial control, including his own essays, the journal took on a critical position of "base materialism."

Bataille's base materialism in *Documents* could be quite literally "base," in terms of social and aesthetic tastes and literal vertical positionality in the world. Bataille's works alone in *Documents* are commentaries accompanied by photographs of severed cow hooves, spiders, and a human big toe. In his article "Le gros orteil"—"The Big Toe"—from the first issue of *Documents* in 1929, Bataille argues for the importance of the literally supporting role of the big toe in differentiating human beings from other animals, affording the "higher" standing of humans in the world, a verticality that, for Bataille, contributes to human cognitive development. Further, the photograph provided in the article as evidence of the importance of the big toe and its role in civilization is not of any famous big toe, but rather of an ordinary and ungroomed, big toe. Indeed, for Bataille, the more aesthetically and often more physically base the object, the more likely it is the material basis for civilization.

Bataille describes the (ig)nobility of the big toe:

> The big toe is the most *human part* of the human body, in the sense that no other element of this body is as differentiated from the corresponding element of the anthropoid ape (chimpanzee, gorilla, orangutan, or gibbon). This is due to the fact that the ape is tree dwelling, whereas man moves on the earth without clinging to branches, having himself become a tree, in other words raising himself straight up in the air like a tree, and all the more beautiful for the correctness of his erection. In addition, the function of the human foot consists in giving a firm foundation to the erection of which man is so proud (the big toe, ceasing to grasp branches is applied to the ground in the same plane as the other toes).
>
> But whatever the role played in the erection by the foot, man, who has a light head, in other words, a head raised to the heavens and heavenly things, sees it as spit, on the pretext that he has his foot in the mud. (Bataille, 1985, p. 20)

Characteristically, as with all infrastructure, the most essential part of the body is often ignored by its users until breakdown:

> Man willingly imagines himself to be like the god Neptune, stilling his own waves, with majesty; nevertheless, the bellowing waves of the viscera, in more or less incessant inflation and upheaval, brusquely put an end to his dignity. Blind, but tranquil and strangely despising his obscure baseness, a given person, ready to call to mind the grandeurs of human history, as when his glance ascends a monument testifying to the grandeur of his nation, is topped in mid-flight by an atrocious pain in his big toe because, though the most noble of animals, he nevertheless has corns on his feet; in other words, he has feet, and these feet independently lead an ignoble life. (Bataille, 1985, p. 22)

Bataille ends "The Big Toe" by challenging the poetic methods of Bretonian surrealism, suggesting that rather than inducing unconscious states in order to see reality, the surrealist poets need only look down at their feet:

> The meaning of this article lies in its insistence on a direct and explicit questioning of *seductiveness*, without taking into account poetic concoctions that are, ultimately, nothing but a diversion (most human beings are naturally feeble and can only abandon themselves to their instincts when in a poetic haze). A return to reality does not imply any new acceptances, but means that one is seduced in a base manner, without transpositions and to the point of screaming, opening his eyes wide: opening them wide, then, before a big toe. (Bataille, 1985, p. 23)

To summarize this chapter, what we see by contrasting the works of Paul Otlet and Georges Bataille is an early twentieth-century drift from an idealistic understanding of documents to a phenomenological one, in both what documents are understood to be and how they appear in the world.

This corresponds to, and in Bataille's case, intersects with, a similar drift in the notion of document in French anthropology (Debaene, 2014). But, what we also see here is how the metaphysics of strong documentarity is inverted, but in such a way as to appear in the empirical world still in the mode of class-type representations. Bataille's base materialism is a philosophy of documentarity grounded in sense, itself taken as a metaphysical category.

In the next chapter I first historically return to a medieval theological context of textual revelation to see the emergence of "literary" styles where the particular illuminates and makes actual the theological universal. Then, I look at another example from earlier twentieth-century documentation theory, where the particular is proposed as absorbed within a strong "scientific" documentarity in documentation practice and through a narrative of scientific revelation. Religious and "literary" texts, here, draw in universal types and themes and give them concrete personages and events in everyday sense and language, while "science" is said to make particulars what they are in documentation practices of ontology, which are understood to be both leading, and in character, a model for, scientific knowledge.

What we see with Otlet and Bataille's works are metaphysical systems of reference and of sense, where particulars are absorbed within transcendental classes and surreal experiences, respectively. Inscriptions are themselves inscribed as reason or experience. What we will encounter in the next chapter are temporal, experiential, movements between universals and particulars within forms of inscription. The universal becomes readable in events and events become readable by documentary science. As different as they are, in both cases, truth is not transcendental, but rather is revealed by performative practices. Truth becomes evident at indexical points of revelation made possible through technologies of informational inscription.

3 Figuring Documentarity

> In the onto-theology of "the West," revelatory unfolding is not just a rhetorical strategy in drama, but rather acts as the chief value index for the meaningfulness of human life and knowledge.

In the previous chapter, we saw the relationship between universal essence and particulars demonstrated in terms of two transcendentalisms: idealist reason and base materialism. These two moments of documentarity, brought into a dialectical relationship, also correspond to two forms of Kantian aesthetics: harmonious beauty and sublime excess. For Otlet, beauty is *illustrated* in depictions of pure reason, and in Bataille's works the sublime is suggested in *photographs* of violence and excess. Their writings also symbolize these aesthetics: a writing of ordered statements and paragraphs, and a writing that at least in content is excessive in sex and violence and might therefore be judged to be pornographic. Despite their opposing epistemologies, aesthetics, and ethics, Otlet's and Bataille's writings share an assumption that empirical particulars are evidence of "deeper" metaphysical truths—reason or "primitive," instinctual drives, respectively. Beings are signs of the general essence or truth of life in their total mode of being. Time plays little role in defining existence between particularity and universality. Even in Bataille's works, where all being is movement and event, all movements and events are then symbols of a transcendental world of drives; "becoming" is an assertion of a type of being. Movement and event are canceled out as real agencies. Bataille's novels display a becoming of excess in eroticism.

Most evidence, whether in literature or the sciences, however, appears through slippages between the real and categories of judgment, as mediated by experience or by methods and technology. What we see as the modern category of realist "fiction," for example, which is the result of the rise of the Western European and British novel of the late eighteenth and early nineteenth centuries, is, as Erich Auerbach (2003) argued, part of a longer tradition or "style" in literature of opposing everyday, experiential accounts of particular, ordinary, individuals against the depiction of persons as belonging to social or other classes, particularly as viewed from an aristocratic conservatism of traditional or eternal values. This former "style" of the everyday I characterize along lines of weak documentarity built on a method of strong sense and the latter as having a strong documentarity built on a method of strong reference based in moral judgment.

"Sense," in the manner that I am using the term, suggests a social or descriptive syntax—which in literature is developed through narrative or through rhetorical figuration or both—rather than dominantly through referential categories. The concept of sense as used here involves experiential doing and undergoing by an agent. It involves directionality and movement, not just through time, but also *as time* for an agent and his or her life and community.

In this chapter, I will prepare for the following chapter, as well (which deals with the modern category of literature as descriptive and performative modes of expression), by presenting two contrary modes by which information as a process of inscription and self-inscription occurs. I take it that the transformation of the sign into an index of meaningful positionality for an agent—that is, as an expressive inscription, rather than as a more "pure" representation or a mere symbolic referent—involves a greater emphasis on sense as a component of its signification.

The two cases that I will discuss in this chapter—aesthetic-religious on the one hand and modern scientific documentation on the other—are expansions and engagements with the juxtaposition of medieval iconography and modern documentation theory that John A. Walsh (2012) performs in his article "Images of God and Friends of God," which discusses medieval religious iconography as indexical signs, using the work of Suzanne Briet (1951, 2006) as an explanatory device. I will argue that the medieval literary and aesthetic modes (at least in the "low" style that we will discuss, reaching at least as far back as the Christian Gospels) and the modern documentation-scientific modes of representation treat the indexical specificity of the particular in opposed manners, but they are both revelatory modes.

In chapter 1, I gave a general philosophical overview between strong and weak documentarity (signs as evidence of classes of universal essences and signs as evidence of particular powers) joined by inscription, and in the previous chapter we saw this mapped onto a theory of signs, still understood as largely referential (even when characterized by a totality of sense and becoming). In this chapter, we get a better understanding of documentarity as inscriptions, though one still tied to figurative uses of signs for theological or documentary-scientific ends. Here, indexes as *figures* of social, cultural, and physical positionality remain inscribed in semiotic documentarity, distinct from powers of self-expression (other than as revelatory expressions). And so, this chapter, despite appeals to empirical particulars, remains overall within the inscriptionality of the sign, rather than ontological powers of self-expression or even an entity's expressions under the constraints of experimentation.

Erich Auerbach's History of Literary Realism

Erich Auerbach's 1946 book *Mimesis: The Representations of Reality in Western Literature* (Auerbach, 2003) is a canonical work that shows the transition from classical to modern literature's depiction of reality from the perspective of rhetorical styles. These styles are of three types: a high style taking the perspective of those from nobler heritages, a comedy or "low style" depicting those of lesser social classes, and most importantly for Auerbach's analysis, a mixed style, made up of realistic depictions of ordinary people. This last develops into modern literary realism. Auerbach's work proposes a historical understanding of literary realism from the perspective of the transition from universal classes of judgment to depictions of particular experience, and so it can help us understand the broader documentary inscriptions that we are tracing out in this chapter and book along a somewhat similar path as Auerbach's work.

From the very beginning of *Mimesis*, Auerbach presents the evolution of the realist style in literature as a break from the method or "style" of assigning particular people and events (particularly of lower classes) to categories or types. In the language of our present book, this is a shift from "strong" documentarity to "weak" documentarity, from reference to sense. Examining this textual history, Auerbach retrospectively reads it in terms of modern literary realism. And, since Auerbach begins his book by discussing biblical and historical narratives along with poetic and narrative epics, the history of the representation of reality according to the modern genre of "literature" *is* this breakage of entities away from *a priori* categories within the classical tradition of narrative, regardless of modern genre identifications. In other words, "literature," as a modern disciplinary or epistemic genre, historically emerges from a style representing powerful particulars that express their identities through experiential relations, whether in history, natural history, poetry, or fiction.

Auerbach (2003) traces the development of such a realism from biblical instances (particularly those in the New Testament Gospels) to the modern novel. Repeatedly, the contrast is between a strong sense of *a priori* categories of knowledge and judgment upon particulars and describing the empirical agency and expressions of particulars in their sensual and experiential lives. Auerbach writes:

> It goes without saying that the stylistic convention of antiquity fails here [i.e., in the Gospels], for the reaction of the casually involved person can only be presented with the highest seriousness. The random fisherman or publican or rich youth, the random Samaritan or adulteress, come from their random everyday circumstances to be immediately confronted with the personality of Jesus; and the reaction of an individual in such a moment is necessarily a matter of profound seriousness, and very often tragic. . . . It goes without saying that, in the New Testament writings, any raising of historical forces to the level of consciousness is totally "unscientific": it clings to the concrete and fails to progress to a systematizing of experience in new concepts. Yet, there is to be observed a spontaneous generation of categories which apply to epochs as well as to states of the inner life and which are much more pliable and dynamic than the categories of Greco-Roman historians. For example, there is the distinction of eras, the eras of law or of sin and the era of grace, faith, and justice; there are the concepts of "love," "power," "spirit" and the like; and even such abstract and static concepts as that of justice have assumed a dialectic mobility (Romans 3: 21ff) which renews them completely.
>
> . . . Surely, the New Testament writings are extremely effective; the tradition of the prophets and the Psalms is alive in them, and in some of them—those written by authors of more or less pronounced Hellenistic culture—we can trace the use of Greek figures of speech. The spirit of rhetoric—a spirit which classified subjects in *genera*, and invested every subject with a specific form of style as the one garment becoming it in virtue of its nature—could not extend its domain to them for the simple reason that their subject would not fit into any of the known genres. (2003, pp. 44–45)

For Auerbach, the history of the representation of reality in Western literature is the story of the emergence of modern realism beginning with what he identifies as the "lower" style of literature—the style beginning with experience, not with categories of judgment—not only in literature proper, but also in religious, historical, and presumably even certain narrative scientific texts. In literature proper, this is characterized by the fictional representation of powerful particulars, independent of *a priori "genera"* or categories. For this reason, modern themes, such as spirit and power, are so important for Auerbach; they represent the innate powers of (and conversely the restrictions on) individual characters to express themselves and give themselves meaning, value, and truth. "Realism" names the power of particulars to name themselves in the narratives that they help create. In literary fiction of the nineteenth century, "realism" refers to the modeling of such particulars as social subjects within story frames.

How then did the experiential narratives of the New Testament become appropriated by the categories of Christian reason and truth in medieval theology? For Auerbach, later Christian church fathers assembled such by figural interpretations of the "realist" narratives, more directly linking prophecies of the Old Testament with New Testament experiential narratives. (Suggestively, "typology" is the name for this hermeneutic technique.) For Auerbach, this resulted in a message that wasn't inherent to the Gospels, but rather, it fitted the needs of the early church in the face of the collapse of the classical order and its desire to impose meaning upon empirical events and their literary recording for institutional ends. The task was documentary in the old, "high-style" sense of classicism, though with an ecclesiastic inflection: to read particular powers as powers of the church's transcendental truths by means of classical rhetorical figuration. Auerbach writes:

> Rigid, narrow, and unproblematic schematization is originally completely alien to the Christian concept of reality. It is true, to be sure, that the rigidifying process is furthered to a considerable degree by the figural interpretation of real events, which, as Christianity became established and spread, grew increasingly influential and which, in its treatment of actual events, dissolved their content of reality, leaving them only their content of meaning. As dogma was established, as the Church's task become more and more a matter of organization, its problem that of winning over peoples completely unprepared and unacquainted with Christian principles, figural interpretation must inevitably become a simple and rigid scheme. But the problem of the process of rigidification as a whole goes deeper;

it is linked to the decline of the culture of antiquity. It is not Christianity which brought about the process of rigidification, but rather Christianity was drawn into it. (2003, pp. 119–120)

Auerbach makes an important point that I will translate into the language of this book: the history of representation follows a general historical progress from strong to weak documentarity, with mixed styles throughout this tendency. The historical quest for the being of beings fluctuates between an emphasis on the powers of types and an emphasis on the powers of particular individuals, with an overall historical tendency toward this latter and toward empirical accounts as the grounds for truth. But typology often comes back to frame the practices of empiricism of whatever eras at their beginnings and their ends, through formal and informal ontologies and measurement parameters, and through practical and technologically produced judgments toward conclusions, as well as throughout in descriptive, "theoretical," accounts. As in literary realism (e.g., narrative as story), the Platonic, that is, the representational, devices of rhetorical closure and philosophical idealism are difficult to escape, even in science or, as we will soon see, in the practices of documentation and its theory.

Figuration and the Indexical Sign

In his article "'Images of God and Friends of God': The holy icon as document" (Walsh, 2012), John A. Walsh applies twentieth-century French documentalist Suzanne Briet's (1951, 2006) concept of the *indice* (indexical sign) in documentation to medieval iconography. Following our discussion of Auerbach's (2003) history of Western literature and the role of representation and figuration in it, a discussion of Walsh's article takes on the importance of showing moments between the universal and the particular where each becomes manifest in the other and showing the rhetorical and experiential mechanisms by which this occurs. Medieval iconography per se is grounded in revelatory metanarratives, though of a strong type of documentarity (iconography as revealing God's truth in the phenomenological world and documentary technique as revealing scientific truth in the phenomenological world). Other literary conventions in the Christian tradition, as Auerbach's book shows, are a weak type of documentarity. In revelatory narratives, however, these are rather relative points on a scale, as it is the manifestation of eternal truths within human reality that constitutes the semiotic universe of belief in this tradition, and literature and art genres merely vary across this scale in order to differently emphasize the causal agencies in life and the epistemic agencies in thought by which revelation occurs.

Despite their radically different time periods and domains of evidence, these two cases, of medieval iconography and Briet's modern documentation, share what Heidegger called an "onto-theological" foundation in both their ontological and historiographic modes of revelation. In Latour's language, inscription and documentary evidence remains at the level of "signs" in both cases: theological and documentary signs. Walsh's work provides an opportunity to consider the role of the indexical sign in documentarity across time periods and genres, from religious reason to a manner of scientific reason, based on documentary class or type construction as acts of evidence and proof.

Walsh's (2012) discussion takes as its starting point a type of figurative art where the nature of the figure is that of the *imago*: the icon. The icon introduces figuration into the medieval philosophical notion of truth as the correspondence between thing and concept, *adaequatio rei et intellectus*. Religious figuration is more than simply a rhetorical or aesthetic function, for it must include experience. The viewer or reader must be brought to a revelation of the meaning of a concept through experience; they must be positioned as a believer of truth by figurative indexing between phenomenological and eternally true worlds. This role of experience distinguishes the functions of literature from those of analytic philosophy in the medieval period. The function of literature, and art more generally, here is to model this discovery of transcendental truth, just like ancient drama led to the discovery of tragedy and comedy in life. Despite their metaphorical functions, icons also involve metonymical trains of experience, particularly when they occur in allegorical rhetoric or symbols. Icons match or condense the phenomenological world to the claims of eternally true concepts and propositions using various metaphorical equivalences and metonymical scales, ending in a one-to-one correspondence between sign and concept.

First, as Walsh points out, as a certain form of allegorical representation, medieval Christian religious icons were created within the context of prototypes and established symbols and beliefs for their creation and interpretation. The hermeneutic interpretability of such was stabilized by a *langue* or structure of interpretative referents. The stability of reference in icons at the level of imagery allows iconographic readings to end in stable interpretations. Walsh writes,

> Artistic invention is not the aim of the icon, which is self-consciously imitating and reproducing from traditional models and prototypes—the Trinity; Mary, Mother of God; the angels; and saints. This faithfulness to earlier models and the prototype provides a stable language, across centuries, for generations of faithful. (2012, p. 186)

Walsh's article points us to an important question in documentary production: the use of figurative language toward producing universal "models and prototypes." As I am arguing throughout this book, the documentary indexes for being vary according to genres and rhetorical strategies, most of all in those worldviews dominated by semiotics. Medieval literature had a repertoire of rhetorical and aesthetic devices and genres that were inherited from classical rhetoric, which as we saw Auerbach suggest, were repurposed for upholding a theological empire, but also served to document the daily external and internal life of ordinary people. Theological arguments subsume the particular within philosophical and theological precedent texts and analytic categories based on assumed universal premises, so that little is left of the empirical other than as argumentative conclusions. And aesthetically, iconography as a tradition subsumed the image of the icon to canonical beliefs and scriptural texts by not only depicting, but also providing through education and the cultural environment, the interpretative meanings for icons. Within the philosophy of iconography, the focus was on the content of the icon as a referent for universal and transcendental meaning.

However, in remembering the experiential aspect of icons, we must also remember their performativity: they were meant to be read. Even the illiterate could "read" an icon; indeed, being broadly readable was central to the icon's function. Icons must also be understood to take on an extensional reference to experience, as well as intensional reference to canonical theological values. They just don't type an event; they must also aesthetically and rhetorically constitute an event by means of the internal correspondence of signs within and between signifying types and regimes. The semiotic economy of signs in and between icons must ultimately find an anchor in experienced life in order for revelation to occur. (Impending death being a convenient place of last resort for the theological, as there are little other grounds for a dying person's hope for continued life than a faith in eternal life.) The document must index a present, as well as an eternal in this lifeworld of documentary evidence. The power of the Holy Trinity, for example, is in the various human personifications of such: in the body of Jesus, the appearance of his mother Mary, and the life paths of the saints. And after his crucifixion, Jesus is not found in the cave, because, by his death, he is now a presence among the living and must always be found there.

This can be seen even more when the distance between the signs or "world" of God and the signs or "world" of human beings are widened by further figurative distance than in the case of singular icons. For example, medieval allegory goes a step further than iconography into the world of empirical life and experiential temporality (both understood as being signs) by setting up correspondences between "the city of man" and "the city of God" that track these correspondences across time and experience. Medieval mystery plays, as live performances, went still further in grounding universals in particular experiences, by embedding the "city" of God within the "city" of human beings. Narratives of mystical experiences embed the universal—the "invisible"—within the particular and visual (or other senses).[1]

In allegory and mystery plays, as well as even in the elevated style of medieval romances, the point of view of God as manifested in the experiential body of Christ remains evident. Truth is revealed as a function of literary form—the Gospels—the eternal "good news" that is active in the present lifeworld. The index for truth is that of God's revelation within the lives of ordinary people, the metonymic appearance of the universal into the particular, and so with this, the awareness that particular human beings have a place in life as a totality. (This would be of particular value to those excluded from power in the political order.) Ultimately, even in the Catholic tradition, the Christian life has meaning through revelation. Without this faith, there is no access to permanence, to eternal life, and so all human life crumbles into finitude and relativism. In the onto-theology of "the West," revelatory unfolding is not just a rhetorical strategy in drama, but rather acts as the chief value index for the meaningfulness of human life and knowledge. "Enlightenment" is temporally progressive and culminating.

Walsh (2012) argues that icons function in networks of indexical signs that link the invisible and the visible, universal essences and visible manifestations of them:

Icons generate complex networks of relationships and indices. They depict relationships among the Persons of the Trinity, the Virgin Mary, the angels, the saints, and the faithful. . . .

The icon manifests and visualizes indexical connections to other documents. These connected documents can be other icons that served as a model or that share the same subject or motif. The referenced artifacts may be scriptural or hagiographic documents that provide a textual representation of the person(s) or events depicted in the icon. The documents indexed by the icon can be the divine or human prototype(s) who are the subject of the image. By providing access to these prototypes, the icon functions as a visible index to the invisible. . . . The invisible is not only referenced by the icon, but is made present in the documentary event, an active participant—with its own envisaging intentionality—in a conversation made possible by the icon. (2012, p. 191)

We can read Walsh above as suggesting not only that icons index the invisible as "documentary event[s]," but that they also perform this transformation of the invisible to the visible by making manifest by images the possibility of the entire medieval Christian worldview. In other words, what is important in the figures of medieval Christianity is not just the relationship of empirical events to biblical characters, events, and moral stories, but what is indexed by the icons most of all is the fact of revelation *in its worldly appearance*. What is historically significant about medieval icons and other figurative forms from a perspective of documentarity is that of revelatory evidence itself as a sign of truth. This occurs in a variety of genres of knowledge in and as "the West," as we will continue to discuss in this chapter and book.

Documentation Theory and the Indexical Sign

As I will now argue, there is also a revelatory understanding of scientific documentation and the practice of collecting and organizing such documents. This understanding of science and documentation has a historical, revelatory, character, not only in its rhetorical and technological manifestations of universal meaning from particular figures, but also in the entire "theology" of its professional destiny: social and epistemic progress through professional documentation (and today, "information" and "data").

Let us return to Walsh's theoretical grounds of his article in the epistemology of the indexical sign or *indice* in Suzanne Briet's *What Is Documentation?*

to further examine how rhetorical figuration functions in Briet's theory of documentation. Briet, we may recall, was not only a librarian at the French Bibliothèque nationale and a founder of French documentation, but she was also a known biographer and theorist of the nineteenth-century French poet Arthur Rimbaud. Despite both these esteemed professional roles, however, neither one ostensibly crossed the other. In Briet's oeuvre, her literary research and research in documentation seem to have no intersections with one another. However, I will ask: can these two regimes be cross-indexed in her work through the figure of the indexical sign, taken as a technology for the transfiguration of knowledge as documentation and of documentation as knowledge?

The role of figuration in philosophical historicism is familiar to anyone acquainted with Hegel's philosophy of history, which gives a giant prefiguration to history by means of the logic of dialectic applied to a very selective and pejorative view of world historical events. In Briet's book, it is not, of course, the logic of dialectic that prefigures history and phenomena, but rather the logic of classificatory evidence—what I've called a "strong" notion of documentarity—and in particular, the figuration of indexicality, a figuration that takes a variety of forms which this current book traces in the form of social agency for evidentiary signs. Ontology, taxonomy, and classification are for Briet indexes of the real. Unlike in Otlet's work, however, where the indexical point is erased in a positivist epistemology of the correspondence of meaning between thing and document, in Briet's book the indexical point is reasserted as a sign created by techniques and institutions of documentation. In Briet's theory of documentation, the phenomenology of essential being lies in the application of the methods and techniques of documentation systems upon individual beings. This, itself, as Briet puts it elsewhere (Briet, 1954) is documentation as the science of science.

In *What Is Documentation?*, Briet writes,

> Is a star a document? Is a pebble rolled by a torrent a document? Is a living animal a document? No. But the photographs and the catalogues of stars, the stones in a museum of mineralogy, and the animals that are cataloged and shown in a zoo, are documents.
>
> In our age of multiple and accelerated broadcasts, the least event, scientific or political, once it has been brought into public knowledge immediately becomes weighted down under a "veil of documents" (Raymond Bayer). Let us admire the

documentary fertility of a simple originary fact: for example, an antelope of a new kind has been encountered in Africa by an explorer who has succeeded in capturing an individual that is then brought back to Europe for our Botanical Garden [*Jardin des Plantes*]. (Briet, 2006, p. 7)

In Briet's book, entities of the world become known as evidence by means of documentary techniques and institutions. These techniques belong to two means: primary or initial documentary techniques of ontologies, taxonomies, and classifications, and then secondary events (lectures, newspaper accounts, etc.) following from these that belong to institutional and popular cultures. So, for example, an animal is named as a new type of antelope through the application of zoological ontologies and taxonomies, and then it is discussed in academic lecture halls as this type of antelope, it is recorded as being the voice of this type of antelope, and its discovery as being this type of antelope is published in newspapers, and so on. Type belongs to a documentary ontology and beings are evidence of the existence of types and are proof of their factual existence. But such truths are products of cultural techniques, foremost, documentation techniques. For Briet, documentation is a *cultural technique*.

Recalling what we read in an earlier chapter of Latour's discussion of bird samples in an aviary (Latour, 1996, p. 27), this transference—indeed, to refer back to religious symbolism, this *transmutation*—of an existent entity into a universal figure of scientific ontologies and taxonomies (and the subsequent constellations of institutional and popular discourses from such), simultaneously makes possible and limits the expressive possibilities of the entity; this individual being must be understood as a *type* within a typology of species and genera. And so, Briet in the beginning of *What Is Documentation?* accepts the standard definition of documents as being evidence of a fact, but she then amends this in order to discuss documents as being indexical signs (*indice*). She writes:

> Latin culture and its heritage have given to the word *document* the meaning of instruction or of proof. RICHELET's dictionary, just as LITTRE's, are two French sources that bear witness to this. A contemporary bibliographer who is concerned about clarity has put forth this brief definition: "A document is a proof in support of a fact."
>
> If one refers to the "official" definitions of the French Union of the Bureaus of Documentation [L'Union Francaise des Organismes de Documentation], one ascertains that the document is defined as: "all bases of materially fixed knowledge, and capable of being used for consultation, study, and proof."

Figuring Documentarity

> This definition has often been countered by linguists and philosophers, who are necessarily infatuated with minutia and logic. Thanks to their analysis of the content of this idea, one can propose here a definition, which may be, at the present time, the most accurate, but is also the most abstract, and thus, the least accessible: "all concrete or symbolic indexical signs [*indice*], preserved or recorded toward the ends of representing, of reconstituting, or of proving a physical or intellectual phenomenon." (2006, p. 7)

In Briet's understanding of documentation, the particular, empirical, antelope is not only symbolically absorbed, but it is also literally captured and enclosed within a cultural institution or institutions. Briet collapses philosophical and documentary notions of ontology and she can do this because the documentary notion *is* philosophical and it is practiced as a cultural technique, not just in the cultures of documentation or philosophy, but as she argues in her book, as Western culture in its social expansion in the world within the mandates of postwar progress and knowledge.

As I have discussed elsewhere (Day, 2006), in Briet's book the animal is captured and brought back to a European collection. This is the professional culture of documentation. But then there is also the cultural destiny of documentation as science, which for Briet rides on the rails of earlier European colonialism, through the dominance of three European languages across the world. In Briet's book, documentation is a professional culture, but it belongs to the metaphysical and political destiny of "the West" as a culture, which in the postwar years is characterized as world development and progress.

Besides the problematic of the documentary appropriation of beings, there is also the problematic that appears of the relationship between figuration and indexicality at the level of the rhetoric and the rhetorical genre of *What Is Documentation?* as a professional manual. For, as we see in the opening paragraphs of Briet's book, the methods and techniques of indexicality are themselves prefigured in Briet's text within the historical destiny of documentation and documentary institutions, as "scientific" institutions.

Documentation, as both a practice and a theory of science for Briet, puts representational limits on signs in their indexical signification in order to form referents of stable and known types, and it is this referential typology that leads to the erasure of figuration by documentary and data institutional processes ("*centres de calcul*," as Latour [1996] uses this term). Figuration and calculation are intractably connected to one another in the technical transformation of signs. In Briet's theory of documentation, through both *a priori* ontologies and taxonomies and then through secondary means of social documentation, the sense of beings is defined and transubstantiated by means of a method and set of techniques whose beginning and end is a referential ontology (and this is true, too, of documentation processes themselves, which are treated as unitary and socially and culturally purposeful).

It is this process of transubstantiation that leads to the transfiguration of the entity as a unification of individual being and universal essence; of turning entities into signs, and these signs into symbols of an essential and universal being. Entities are allegorized as signs of universal truth, emerging through processes of scientific revelation, led by ontological naming, whereby the entity gains its importance and value for truth by representing something other than its own particularity, a mode of generalized being that transcends particular entities and that appears through vigorous methods and techniques. The entity has been changed by naming, and again by not just being an inscription, but rather a symbol, of being revealed and made manifest through "rigorous" thought or reason. Briet's book describes the transubstantiation and transfiguration of an entity by science—"science" understood as a process of ontological naming mediated by documentation techniques, methods, and institutions. That is, "science" itself understood as documentation or, as Otlet had it, as bibliographic representation. European documentation forms a direct link between the representational tradition in bibliography and the notion of information in modern information science; books as containers of facts, documents as containers of facts, and the contents of those containers—the facts—being information. Science, in this sense, is synonymous with what Heidegger called the "ontotheological" tradition of Western metaphysics. It attempts to reveal the *logos* of all that is.

In this chapter, we have looked at the figuration of entities as evidentiary signs of truth within a medieval and a modern context. We have seen experience and particular empirical entities subsumed within semiotics of signs operating as revelatory mechanisms for truth. In the next chapter, we look at entities as represented social entities within the genre of modern literary realism. We will also look at the critique of representational realism by the modern literary and aesthetic avant-garde.

4 Documentarity and the Modern Genre of "Literature"

The inhabitants of bourgeois worlds are premised and defined by institutions and fantasies of second-hand knowledge—evidentiary texts, or "documents," both of "fact" and "fiction"—including the knowledge of their own personal and social psychologies. Bourgeois identity and life is mediated throughout, in both its "interior" (the supposedly personal psychological "inside") and "exterior" (the supposedly social psychological "outside") aspects by documents. Its primary psychological indexes are built through documentation and documentarity.

At least since Western European Romanticism of the late eighteenth and early nineteenth centuries, the written, spoken, and visual arts have been seen as the domain of human cultural expression, particularly the expressions of the rather recent conception of the modern individual self. With written texts, this domain has traditionally been thought to belong to literature. Popularly, literature is thought to be expressive of personal emotions or "feelings" and the depiction of fictional events and poetic manners of expression.

However, "literature"—more literally as it were—means that which involves writing generally. Just as in psychology the dividing line between emotive and cognitive expressions can be problematic, so, too, the division between literature and documentation becomes problematic, not least when we examine both as being "informational."

If we view the specific genre of "literature" as a more or less modern phenomenon—for example, if we ask how and why there came to be "literature" as a category of information distinct from documentation—then we may begin to see literature as a reaction since the beginnings of modernity to the dominance of documentarity, as well as being part of the emergence of a new type of document or information that is grounded in the depiction of beings as broadly powerful particulars.

As we have discussed, though there have been styles that marked fiction as distinct from philosophy before the modern period, "literature" (particularly having to do with the self as a powerful particular and with the extended depictions of ordinary persons in realistic manners) is a modern genre. And while there were earlier works, such as the well-known example of Laurence Sterne's eighteenth-century novel, *The Life and Opinions of Tristram Shandy, Gentleman*, which "showed the devices" of their construction, only with the modern avant-garde did this become a more generalized approach that also took "literature" (as a realist, representational, form) into its critical view. While the popular view is that it was the modern sciences that first separated themselves from a previous prescientific form of writing and investigation that was symbolic in character, I would suggest that it was modern literature, too, as a novelistic realm, that separated itself from such a symbol-based writing (save for modern literary symbolism, such as the Silver Age of Russian literature), and that it did so by mimicking the empirical sciences' concerns with the particular, while also claiming to go beyond their abilities and domains of knowledge by literature's ability to "organically" or "holistically" depict the emotional and social underpinnings of human beings and their relations to one another. "Literature," as much as modern science, constituted a break from what came before, in so far that in both cases it was the symbol as the basis for meaning, which was supposedly distanced from and replaced by realist representation, and in the avant-garde, by performance and materialism.

While there certainly were lyric poets and there were prose perspectives in the first and third person before the modern period, the central concern of the self as an agent of feelings, particularly as one undergoing changes in character due to the social environment, became a unique hallmark of literature in the modern period and still today. Wordsworth's diction as representative of not only his own voice, but also the expressions of the "ordinary" rural man or woman, Whitman's similar concerns in his "Song of Myself," Baudelaire and Rimbaud's self-concerns in regard to social modernity, and the focus on the self in Beat poetry, Anglo-American "confessional" poetry, and New American Poetry of the 1960s—all these and so many more examples would have been impossible previous to the rise of the conception of the self in the seventeenth century, as a powerful particular with (as we will discuss in a later chapter) "inalienable" rights of expression and being. In romantic and broadly understood "modern" literature, an ancient literature of what Auerbach (2003) calls "high style" (depicting the aristocratic self as undone by hubris and fate—ancient tragedy through Shakespeare), or conversely, a "low style" (characterized by rabble or clever underdogs—Latin literature and up through picaresque novels), in other words, a literature of styles based on character types, is largely replaced by realist depictions of selves within social conditions. Likewise, French realism in the nineteenth-century novel took what was largely a woman's form of high-style literature in the previous century, one that recirculated medieval romances, and turned this form on its head, situating the reader in the current or near-current moment and sometimes parodying the effects of "high-style" romances on more ordinary readers of the emerging middle class (e.g., in the case of Gustave Flaubert's *Madame Bovary*, which we will soon examine).

The argument that I will pose in this chapter is that modern literature develops as a counter-discourse to the judgmental categories of the "high-style" literature preceding it, the latter of which followed a strong form of documentarity. And I suggest that it does this through not only a greater concern with ontological particulars, but also out of greatly expanded and intensified acts of depicted and performed sense associated with these particulars. In this, it gives a descriptive focus similar to the rise of experimentalism at the time, although in two forms: that of modeling via fiction and that of performance and formal critique and innovation. Thus, I propose three categories for modern literature and for art ("literature"—that is, aesthetic inscriptions more generally), in order to discuss them in relation to the "strong," *a priori* documentarity that came before and accompanied them in both aesthetic and non-aesthetic practices:

1. as the expansion and intensification of sense through temporally expanded fictional description and powerful particulars as characters (e.g., the modern realist novel);
2. as the intensification of sense in empirical affects (i.e., the modern avant-garde as characterized by a privileging of sensation against normative cognitive representation or "knowledge"; e.g., Dada shock value, sound poetry, and up through 1960–70s, performance art and "happenings");
3. as the "deconstructive" (loosely phrased) revealing of the devices and the making strange of everyday discursive and also literary expressions that claim realist or naturalist representation (e.g., formalist and constructivist traditions from the time of the Soviet avant-garde to contemporary works in "Language Writing" [sometimes called "Language Poetry"]).

Our inquiry in this chapter is accompanied by questions (which we can only touch on in this book) about the relationship of "literature" and "information" today, such as: What is the status of the text in the modern documentary episteme of the "information age"? How does literature attempt to remain specific in the midst of the dominance of the current documentary episteme of information and data? How does it react to this dominance? How is "information" understood as distinct from modern documentation, and how does this have analogies or precedence in literature? What is the status of literature and "the humanities" (including the "digital humanities") now that so much of our social world is textually mediated as stories and through poetic devices? And, do documentation and literature from the nineteenth century and into the twenty-first century form two poles of the modern any longer, a "modern" that has as its central idea that of information as evidence and proof (i.e., modern documentarity)?

In terms of the above third category, we must be aware that the performative critique of documentation through formal means has occurred not only in art and literature, but also in certain areas of philosophy, literary criticism, and even in the sciences in the twentieth century and into the twenty-first century (i.e., in anthropology and ethnology [Debaene, 2014]) through a critique and performance of the notion of "the text," as well as through the integration of literary style or textual analysis within these domains. The critical notion of "the text," which pinnacled in French philosophy of the 1960s and '70s and subsequently Anglo-American theory, takes on a particular character that poses it against not only documents as evidence, but also philosophy as a metaphysics of documentary evidence. The explicit literary character of some philosophy during the late nineteenth and twentieth centuries (e.g., Nietzsche's *Thus Spoke Zarathustra*, as well as his use of aphoristic forms in his other books; Heidegger's neologisms; Derrida's blurring of literary and philosophical styles) adds to performative critiques of strong documentarity during late modernity.

I can claim little originality for some of the historical material given in this chapter. Much of it is well known and canonical. However, this is partly the point. How such became or becomes literature, as a genre, or has become canonical as art or poetics or literary movements, remains an issue that must be accounted for not only in terms of the history and sociology of the works and the biographies of literary writers and artists, but also in terms of larger sociological, conceptual, and historical explanations, not least those involving documentation, and today, the question of information or documentarity in regard to the text.

What is perhaps original in this chapter is the suggestion that the modern category of literature has appeared in different ways as both critical and intensifying to documentation, as part of an empirical response to strong documentarity (reference) and in support of weak documentarity (sense). To investigate "information in the humanities" doesn't just mean to investigate bibliographic and other such resources at the service of humanistic and artistic practices, but to investigate the humanities and arts in terms of their critical and their supportive relationships to modern conceptions of information and to documentarity more fully.[1]

Literary Realism

Nineteenth-century naturalism as a type of realism, influenced by the social sciences of the day, focused on an aesthetic intensification of behavioral qualities according to social types (e.g., Zola's works). Realism more generally, though, was committed to representing a more phenomenologically broad and more emotionally intense range of social behavior than could be afforded by the methods of the social sciences. Through fictional works, imaginative and psychological realms of particular persons and social spaces could be iconically modeled, in a past, present, and even in an imagined future.

Literary realism stresses the depiction of the psychologically and socially caused affective relations of human characters, while modeling their behaviors for the reader to identify with. It attempts to go beyond behavioral taxonomies by displaying the confused motivations and intentions of characters as selves, as well as social types of persons. Even in third-person narratives, the point of view is the character as an agent of expressive powers, whether as an agent of free will or as a reactive victim of circumstance.

I will begin my analysis of modern literature by discussing nineteenth-century realism as a literary genre that forefronts the self-evidentiary expressions of modeled characters in story forms. My example will be Gustave Flaubert's famous 1856 novel *Madame Bovary*, composed of narrative representations built from the viewpoint of character-agents. Character, in the moral sense of the word, is the "virtual" or essential properties of selves that are actualized in situations and events. In the realist novel, as in real life, we hypothesize character—or a self—out of the regularity of complex expressions from an agent. The complexities and the contradictions of self are what separates the realism of Flaubert's *Madame Bovary* from the satirical characters of his later *Bouvard et Pécuchet* or the naturalism of social types in Zola's works, for example. The modern novel largely functions as the depiction of character in an extended story frame. Both elements—characters as selves and stories as complex, moralistic, allegories or models for lived life—are important elements in the genre form of the modern realist novel.

Madame Bovary

Gustave Flaubert's celebrated novel of the mid-nineteenth century, *Madame Bovary*, continues the turn of realism from a stronger documentarity of literature (Auerbach's "high style") to a "weaker" one. In Flaubert's type of realism in *Madame Bovary*, it is agents as powerful particulars that shape their destiny, rather than just their social types. While social forces are powerfully portrayed in *Madame Bovary*, and while the secondary characters conform strongly to the social types of the time, the central character of Emma Bovary displays a particularly powerful will in the midst of being shaped by social forces in the present and from the past. Her fate is a product of this will, even as it becomes recommodified by literary types, and personalities (e.g., the character of Rodolphe) of previous centuries. As distinct from the "high style" of ancient depictions up through the early modern period, Emma is an everyday, bourgeois character who still today bourgeois readers can identify with.

To understand the emergence of literary realism from a documentary influenced naturalism, which both came after it in literature and to some extent preceded it more broadly in writing (Auerbach's "high style"), we may step back and look at an extended passage from Honoré de Balzac's 1842 preface to his *The Human Comedy*. There, Balzac presents the emergence of human social types as analogues to animal evolution in nature, while adding the complexity of the human self. With, on the one hand, eighteenth-century natural philosophy and, on the other hand, subsequent Darwinian influenced biology and the emerging social sciences in the background, Balzac writes:

> The idea of *The Human Comedy* was at first as a dream to me, one of those impossible projects which we caress and then let fly; a chimera that gives us a glimpse

of its smiling woman's face, and forthwith spreads its wings and returns to a heavenly realm of phantasy. But this chimera, like many another, has become a reality; has its behests, its tyranny, which must be obeyed.

The idea originated in a comparison between Humanity and Animality.

It is a mistake to suppose that the great dispute which has lately made a stir, between Cuvier and Geoffroy Saint-Hilaire, arose from a scientific innovation. Unity of structure, under other names, had occupied the greatest minds during the two previous centuries. As we read the extraordinary writings of the mystics who studied the sciences in their relation to infinity, such as Swedenborg, Saint-Martin, and others, and the works of the greatest authors on Natural History—Leibnitz, Buffon, Charles Bonnet, etc., we detect in the *monads* of Leibnitz, in the *organic molecules* of Buffon, in the *vegetative force* of Needham, in the correlation of similar organs of Charles Bonnet—who in 1760 was so bold as to write, "Animals vegetate as plants do"—we detect, I say, the rudiments of the great law of Self for Self, which lies at the root of *Unity of Plan*. There is but one Animal. The Creator works on a single model for every organized being. "The Animal" is elementary, and takes its external form, or, to be accurate, the differences in its form, from the environment in which it is obliged to develop. Zoological species are the result of these differences. The announcement and defense of this system, which is indeed in harmony with our preconceived ideas of Divine Power, will be the eternal glory of Geoffroy Saint-Hilaire, Cuvier's victorious opponent on this point of higher science, whose triumph was hailed by Goethe in the last article he wrote.

I, for my part, convinced of this scheme of nature long before the discussion to which it has given rise, perceived that in this respect society resembled nature. For does not society modify Man, according to the conditions in which he lives and acts, into men as manifold as the species in Zoology? The differences between a soldier, an artisan, a man of business, a lawyer, an idler, a student, a statesman, a merchant, a sailor, a poet, a beggar, a priest, are as great, though not so easy to define, as those between the wolf, the lion, the ass, the crow, the shark, the seal, the sheep, etc. Thus, social species have always existed, and will always exist, just as there are zoological species. If Buffon could produce a magnificent work by attempting to represent in a book the whole realm of zoology, was there not room for a work of the same kind on society? But the limits set by nature to the variations of animals have no existence in society. When Buffon describes the lion, he dismisses the lioness with a few phrases; but in society a wife is not always the female of the male. There may be two perfectly dissimilar beings in one household. The wife of a shopkeeper is sometimes worthy of a prince, and the wife of a prince is often worthless compared with the wife of an artisan. The social state has freaks which Nature does not allow herself; it is nature *plus* society. The description of social species would thus be at least double that of animal species, merely in view of the two sexes. Then, among animals the drama is limited; there is scarcely any confusion; they turn and rend each other—that is all. Men, too, rend each other; but their greater or less intelligence makes the struggle far more complicated. Though some savants do not yet admit that the animal nature flows into human nature through an immense tide of life, the grocer certainly becomes a peer, and the noble sometimes sinks to the lowest social grade. Again, Buffon found that life was extremely simple among animals. Animals have little property, and neither arts nor sciences; while man, by a law that has yet to be sought, has a tendency to express his culture, his thoughts, and his life in everything he appropriates to his use. Though Leuwenhoek, Swammerdam, Spallanzani, Reaumur, Charles Bonnet, Muller, Haller and other patient investigators have shown us how interesting are the habits of animals, those of each kind, are, at least to our eyes, always and in every age alike; whereas the dress, the manners, the speech, the dwelling of a prince, a banker, an artist, a citizen, a priest, and a pauper are absolutely unlike, and change with every phase of civilization. . . .

. . . Though dazzled, so to speak, by Walter Scott's amazing fertility, always himself and always original, I did not despair, for I found the source of his genius in the infinite variety of human nature. Chance is the greatest romancer in the world; we have only to study it. French society would be the real author; I should only be the secretary. By drawing up an inventory of vices and virtues, by collecting the chief facts of the passions, by depicting characters, by choosing the principal incidents of social life, by composing types out of a combination of homogeneous characteristics, I might perhaps succeed in writing the history which so many historians have neglected: that of Manners. By patience and perseverance, I might produce for France in the nineteenth century the book which we must all regret that Rome, Athens, Tyre, Memphis, Persia, and India have not bequeathed to us; that history of their social life which, prompted by the Abbe Barthelemy, Monteil patiently and steadily tried to write for the Middle Ages, but in an unattractive form. (Balzac, 1901, pp. 2–5)

For Balzac, human social types exist, but they are variable and contested by individual character, complexities of culture or "manners" and chance. Characters in realist fiction, as individuals in real life, are made up of both selves (composed of unique collections of potential skills and expressions) and persons (composed of actions that conform to socially accepted roles and rules for behavior—i.e., to moral orders). The variances or deviations between self and person and how an individual negotiates this as a manifestation of their ontological and political "freedom" is one of the characteristics that make up realist works in this period and our own. Unsurprisingly, the realist novel arose in post-Enlightenment societies, when new social and psychological norms for "free" individuality spread, emphasizing the singularity of the bourgeois self, even in the midst of its intense social and cultural documentary construction and self-construction. We can see this in *Madame Bovary*, where the characters of a small country town appropriate as self-identities bourgeois social types taken from national and international cultural materials, such as newspapers and novels. They then contest and modify these person-types in their local circumstances and according to each, his or her, own self character, within an overall political narrative of social progress and individual self-development.

Through reading mass media materials, the characters appropriate imaginations of their own individual being, which, in the case of Emma Bovary as a woman, results in her imaginations of romance according to aristocratic types. In *Madame Bovary*, we have a depiction of the historical transition from an aristocratic moral code to a bourgeois one, the latter which has appropriated some of the former's figures, though always mediated by the need for a wage, and through it, personal and social progress (at least economically). Tragically, Emma takes her cad of a lover, the aristocrat Rodolphe, to be a model for the freedom to ignore what she sees as boring and hypocritical bourgeois social rules and roles, while she also looks down upon the rural peasant moral orders. This turns out tragically, because the freedoms of the aristocracy are not shared by the bourgeoisie, whose very existence and social advancement are dependent on social types and moral orders ideologically generated and maintained by newspapers, the church, education, and other such—largely documentarily mediated—cultural institutions.

Emma's husband Charles was not a very enthusiastic student in his youth, though by the time he marries Emma, he has become—thanks to his mother's overbearing force—a doctor, and he lives his life with Emma as a caring, but for her, dreadfully boring, husband, following the roles, rules, and social privileges of a small-town doctor. Flaubert depicts Charles, along with the other bourgeois men in the novel (to appropriate the English translation of the Austrian writer Robert Musil's novel title), as a "man without qualities." Charles accepts his social rules and roles, which have given him, despite his demonstrated incompetence, a certain level of bourgeois and professional respectability in the region, which Flaubert skews unremittingly, both through Emma's eyes and through dry vignettes.

Probably one of the most tragic-comic illustrations of Charles's pretentions as a doctor is the scene of his botched attempt to correct a laborer's club foot, resulting in gangrene, the man's foot being painfully amputated, and Emma's further disdain for Charles based on her concern for damage to their social status if his reputation should fall. Flaubert's description of

Charles's approach toward the amputation and Emma's disdain for his clumsy efforts beautifully portrays the complicated manners by which social identities—derived from popular and professional documents in the present and from the past—position the characters as social actors and drive their desires and relationships. Emma takes Charles's failure at the operation as one more failure of his character; he is, in her eyes, a buffoon, a caricature of a lover and of a doctor.[2]

It should be stressed, before quoting this section of *Madame Bovary*, that one of the hallmarks of documentarity in modernity is the extensiveness by which mass media creates identity types and how these shape both individual and social psychologies and politics, most of all in bourgeois social spaces where experience is heavily mediated by documentary, that is, "second-hand" knowledge (Wilson, 1983). In the bourgeois sphere, one advances by the attainment of technical skills and cultural awards, through education, institutions, and its documents; in other words, one socially advances through using and being granted evidentiary texts. Social powers are not inherited, as with the aristocracy, but rather must be striven for as class achievements, and this requires earned recognition in the public, and indeed, the private spheres. Personality is constructed and publicly expressed through writing, or more broadly understood, inscriptions. Documentarity, through the literal processes of documentational learning and attainment (documentality), accompanies moral psychology during the modern age. One earns and gives documentary evidence about who one is. One publishes texts and one "earns" (or otherwise receives) diplomas, and these constitute the professional and moral character of a person for the rest of his or her life. Public documents have the aura of creating permanent moral and psychological identity—of giving a trustworthy permanence to identity against time, interpretation, illness, cognitive variance, and the emotional flux of human relationships.

Flaubert writes in *Madame Bovary*:

> He [Charles] had recently read a eulogy on a new method for curing club-foot, and as he was a partisan of progress, he conceived the patriotic idea that Yonville, in order to keep to the fore, ought to have some operations for strephopody or club-foot.
>
> "For," said he to Emma, "what risk is there? See" (and he enumerated on his fingers the advantages of the attempt), "success, almost certain relief and beautifying the patient, celebrity acquired by the operator. Why, for example, should

not your husband relieve poor Hippolyte of the 'Lion d'Or'? Note that he would not fail to tell about his cure to all the travelers, and then" (Homais lowered his voice and looked around him), "who is to prevent me from sending a short paragraph on the subject to the paper? Eh! goodness me! an article gets about; it is talked of; it ends by making a snowball! And who knows? who knows?"

In fact, Bovary might succeed. Nothing proved to Emma that he was not clever; and what a satisfaction for her to have urged him to a step by which his reputation and fortune would be increased! She only wished to lean on something more solid than love.

Charles, urged by the chemist and by her, allowed himself to be persuaded. He sent to Rouen for Dr. Duval's volume, and every evening, holding his head between both hands, plunged into the reading of it.

While he was studying equinus, varus, and valgus, that is to say, *katastrephopody, endostrephopody,* and *exostrephopody* (or better, the various turnings of the foot downwards, inwards, and outwards, with the *hypostrephopody* and *anastrephopody*), otherwise torsion downwards and upwards, Monsieur Homais, with all sorts of arguments, was exhorting the lad at the inn to submit to the operation.

. . . The poor fellow gave way, for it was like a conspiracy. Binet, who never interfered with other people's business, Madame Lefrançois, Artémise, the neighbors, even the mayor, Monsieur Tuvache—every one persuaded him, lectured him, shamed him; but what finally decided him was that it would cost him nothing. Bovary even undertook to provide the machine for the operation. This generosity was an idea of Emma's, and Charles consented to it, thinking in his heart of hearts that his wife was an angel.

So, by the advice of the chemist, and after three fresh starts, he had a kind of box made by the carpenter, with the aid of the locksmith, that weighed about eight pounds, in which iron, wood, sheet-iron, leather, screws, and nuts had not been spared.

But to know which of Hippolyte's tendons to cut, it was necessary first of all to find out what kind of club-foot he had.

He had a foot forming almost a straight line with the leg, which, however, did not prevent it from being turned in, so that it was an equinus together with something of a varus, or else a slight varus with a strong tendency to equinus. But with this equinus, wide in foot like a horse's hoof, with rugose skin, dry tendons, and large toes, on which the black nails looked as if made of iron, the club-foot ran about like a deer from morn till night. He was constantly to be seen on the Place, jumping around the carts, thrusting his limping foot forward. He seemed even stronger on that leg than the other. By dint of hard service it had acquired, as it were, moral qualities of patience and of energy; and when he was doing some heavy work, he stood on it in preference to its fellow.

Now, as it was an equinus, it was necessary to cut the tendon Achilles, and, if need were, the anterior tibial muscle could be seen to afterwards for getting rid of the varus; for the doctor did not dare to risk both operations at once; he was even trembling already for fear of injuring some important region that he did not know.

Neither Ambrose Paré, applying for the first time since Celsus, after an interval of fifteen centuries, a ligature to an artery, nor Dupuytren, about to open an abscess in the brain, nor Gensoul when he first took away the superior maxilla, had hearts that trembled, hands that shook, minds so strained as had the doctor when he approached Hippolyte, his tenotome between his fingers. And, as at hospitals, nearby on a table lay a heap of lint, with waxed thread, many bandages—a pyramid of bandages—every bandage to be found at the chemist's. It was Monsieur Homais who since morning had been organising all these preparations, as much to dazzle the multitude as to keep up his illusions. Charles pierced the skin; a dry crackling was heard. The tendon was cut, the operation over. Hippolyte could not get over his surprise, but bent over Bovary's hands to cover them with kisses.

"Come, be calm," said the chemist; "later you will show your gratitude to your benefactor."

And he went down to tell the result to five or six inquirers who were waiting in the yard, and who fancied that Hippolyte would reappear walking properly. Then Charles, having buckled his patient into the machine, went home, where Emma, all anxiety, awaited him at the door. She threw herself on his neck: they sat down to table; he ate much, and at dessert he even wished to take a cup of coffee, a luxury he permitted himself only on Sundays when there was company. (Flaubert, 1904, pp. 217–221)

The operation, though, ends in failure, as the foot badly gangrenes. Flaubert later describes the corrective operation by a renowned Dr. Canivet from Paris, which Charles and Emma hear taking place at a distance by the fact of Hippolyte's screams. Charles and Emma's reaction to this, like all the other characters in the town, is self-centered, focused on those identities that they hold of themselves courtesy of the popular media and the popular and learned "opinions" that have bequeathed them; in the case of the Bovarys' that of a respected professional man of medicine and his wife:

Then, without any consideration for Hippolyte, who was sweating with agony between his sheets, these gentlemen entered into a conversation, in which the chemist compared the coolness of a surgeon to that of a general; and this comparison was pleasing to Canivet, who launched out on the exigencies of his art. He looked upon it as a sacred office, although the ordinary practitioners dishonoured it. At last, coming back to the patient, he examined the bandages brought by Homais, the same that had appeared for the club-foot, and asked for someone to hold the limb for him. Lestiboudois was sent for, and Monsieur Canivet having turned up his sleeves, passed into the billiard-room, while the chemist stayed

Documentarity and Modern Literature

with Artémise and the landlady, both whiter than their aprons, and with ears strained towards the door.

Bovary during this time did not dare to stir from his house. He kept downstairs in the sitting-room by the side of the fireless chimney, his chin on his breast, his hands clasped, his eyes staring. "What a mishap!" he thought, "what a mishap!" Perhaps, after all, he had made some slip. He thought it over, but could hit upon nothing. But the most famous surgeons also made mistakes; and that is what no one would ever believe! People, on the contrary, would laugh, jeer! It would spread as far as Forges, as Neufchâtel, as Rouen, everywhere! Who could say if his colleagues would not write against him. Polemics would ensue; he would have to answer in the papers. Hippolyte might even prosecute him. He saw himself dishonored, ruined, lost; and his imagination, assailed by a world of hypotheses, tossed amongst them like an empty cask borne by the sea and floating upon the waves.

Emma, opposite, watched him; she did not share his humiliation; she felt another—that of having supposed such a man was worth anything. As if twenty times already she had not sufficiently perceived his mediocrity. (1904, pp. 229–230)

Emma and Charles and many of the other, bourgeois characters inhabit a world heavily mediated by second-hand (i.e., documentary) knowledge. Just as bibliographies and other library indexes position and point to books within collections of library texts, and so as they also position persons as types of users in regard to them, media spheres index documentary materials and bring people as social types—"users"—into relation with such information and thus to themselves as identities, as well (Day, 2014). Unlike peasant illiterates and near illiterates, and unlike the aristocrats who could travel or have foreign guests, the middle class (then and now) largely gained privilege, enlightenment, knowledge about the world, and power secondhand, through books, newspapers, public and institutional documents, and other documentary information. At the same time, lack of formal education and reading among the peasantry and proletariat were (and still are) seen by the bourgeoisie (today, middle or even upper-working) class as a sign of the lower classes' ignorance and lack of social progress, even when their suffering is a result of trusting the assumed knowledge held by the bourgeoisie. The "ignorance" of the lower classes supposedly is not only of the contents of such documents, but also of the "fact" that such sources of knowledge bring power. This is the trust that such lower classes supposedly lack, which is the contract that makes up the bourgeois state, understood as "democratic" and "progressive." That the documents that the bourgeoisie have access to are usually mediated by corporate or public spheres controlled by wealth, and that the bourgeoisie often lack direct access to research or the ability to understand such documents (e.g., medical research), is a lacuna that in itself constitutes their very faith in institutional documents, institutions, and professions. Their faith in their own ability to progress and their faith in progressive nationhood or "civilization," that is, their faith that knowledge is power, are precisely rooted in their ignorance. Their knowledge is not even secondhand in most cases, but further down the scale. The entirety of bourgeois existence and their trust in the state lies in a faith in documentary mechanisms largely beyond their immediate knowledge.

If the aristocracy exists through their own otherworldly fantasies and self-satisfaction, the elite through knowledge of the sociocultural mechanisms and facts of power that keep them in power, the poor through local "street smarts," and the peasants through tradition and immediate values (especially if they are illiterate—e.g., children instead of speculative financial investments as insurance for the future; successful harvests and food on the table instead of long term, speculative threats; warm and long lasting clothes instead of fashion[3]), then the bourgeoisie existed in the early nineteenth century and still exist today through the mediation of documents and other texts, which today include all types of media, including their own imaginations of actively performing all the above classes in bourgeois fantasies of personal progress, personal success, manual labor, and wealth and leisure (e.g., going back to the land; affordable fashion; supervisory roles in jobs; voting; and land ownership, vacations, and moments of relaxation and luxury). The inhabitants of bourgeois worlds are premised and defined by institutions and fantasies of secondhand (or thirdhand) knowledge—evidentiary texts, or "documents," both of "fact" and "fiction"—including the knowledge of their own personal and social psychologies. Bourgeois identity and life is mediated throughout, in both its "interior" (the supposedly personal psychological "inside") and "exterior" (the supposedly social psychological "outside") aspects by documents. Its primary psychological indexes are built through documentation and documentarity.

However, as shown in *Madame Bovary*, even among the bourgeoisie, there was and continues to be a hierarchy of knowledge that splits along lines of high-culture and "morally virtuous" canonical journalistic, philosophical, and scientific literature on the one hand, and that of other types of "lower" literature (fictional and romantic) on the other. These latter were the documents of life, however, that became much of the modern genre of "literature," based on the supposed difference between evidence-in-fact and evidence-in-fiction.

In France, women novelists such as Madame de Tencin (1682–1749) and her *Memoires du Comte de Comminge* and Madeleine de Scudéry (1607–1701) and her novels, and the many similar authors at the time and those that came after them, provided the romance, dialogue, and beyond all else, the rules and roles necessary for a woman's social advancement (DeJean, 1991), all of which were lacking in Emma Bovary's sheltered bourgeois house and village existence. (One of the striking elements in Flaubert's novel is Emma's almost total lack of female companionship, save for her nanny and servants, whom, of course, Emma shows little empathy with or sympathy toward.) The fancifulness of the novels and their settings provided "as-if" models for women, particularly perhaps those cut off from other women by class or domestic isolation.

In *Madame Bovary*, Flaubert ruthlessly displays the limits and powers of this secondhand knowledge and its importance in creating personal and social identity for the characters, while also showing that the new genre of "literature" provided a rare textual outlet for the emotional aspects of their lives. Below, for example, is Flaubert's commentary on Leon—another, earlier, love interest for Emma just after her marriage to Charles—in a scene later in the book depicting Leon's renunciation of romantic passions. Literature is shown as a sociological reaction to an organized, documentary world ("senior clerk," "notary"), but such literature is, for bourgeois culture, also filled with stereotypes and tropes ("immense passions," "heroic enterprises," "oriental princesses"), though of a supposedly "escapist" form, and so is inappropriate for a young man attaining a professional or semiprofessional social status. Flaubert writes,

> At last Leon swore he would not see Emma again, and he reproached himself with not having kept his word, considering all the worry and lectures this woman might still draw down upon him, without reckoning the jokes made by his companions as they sat around the stove in the morning. Besides, he was soon to be head clerk; it was time to settle down. So, he gave up his flute, exalted sentiments, and poetry; for every bourgeois in the flush of his youth, were it but for a day, a moment, has believed himself capable of immense passions, of lofty enterprises. The most mediocre libertine has dreamed of sultanas; every notary bears within him the debris of a poet. (1904, p. 77)

Here, Flaubert tell us that in the social development of men the emotional and literary world of women ("passions") must be renounced; fiction gives way to documents, evidence of desire to facts. The sexual division of labor takes place along lines of private and public spaces for evidence, with sexual desire as the dividing line, for what is considered essential and also practical. Sexual desire—at least understood as fulfilled or unfulfilled passions—demarcates sexual identity, while also confining the two sexes to private and public spheres that follow this same demarcation.

While characters such as Rodolphe could live the last vestiges of the aristocracy and enact the heroic or "high style" of literary character (and also abandon it at will, as he abandons Emma), it was up to women like Emma to evoke these aristocratic privileges of "passion" and natural right against emerging bourgeois values, earned rights, and technological progress. In truth, however, both Charles and Emma are situated in moral and social classes of earned, not aristocratically inherited, rights. They cannot escape their bourgeois class assignment because to do so would be to abandon the very means toward class advancement that they dream of and to abandon the moral rules and roles, without which they would fall into the working class and peasantry. This is the reason why all the main characters other than Rodolphe (who is a self-absorbed hypocrite) in *Madame Bovary* can be nothing other than essentially self-parodic: in whatever way they reach out for freedom, it is in a way that has already been mediated for them by means of bourgeois culture itself and the political state that claims to represent it.

The aristocracy lives in a world of surplus fantasy, but the bourgeoisie live in a world of *necessary* imaginaries, mediated by documentary evidence of all sorts. Documents in all media types make up much of the symbolic capital of bourgeois life. All their striving for an aristocratic wealth is already foreclosed in the very means of getting it (for to be aristocratic means to inherit that surplus—not just financial, but moral). All the bourgeois institutions that educate and certify the bourgeoisie are, paradoxically, the means to both their desired freedom and their class enslavement. They seek to earn by the wage the freedom to live without it. They try to live the "free" sexual and financial lives of the aristocrats and they end up ruined and morally condemned. Emma and Charles and their ilk demonstrate that their social class and the modern psychology of individuality associated with it arise together: the individual is free insofar as he or she occupies a social type that is seen as free, and so, such persons can never be free in the same way as those born to inherited wealth (nor even to those who are illiterate). Mediation—epistemic, sociological, psychological, and in sum, ontic—in what I've called the "modern documentary tradition" (Day, 2014)—is deeply tied to class position and the documentary means for it.

Emma is locked in a struggle for freedom against the governing mechanisms of taste and social rules and roles that inscribe her, but she wages her battle through the same inscribing mechanisms that limit her freedom. Her battle is a flight against capture done through inscribed sexual lines, through genre forms that both capture and intensify her emotional needs and inscribe her as a bourgeois woman and wife of her time. The novels that she reads provide her with the "evidence" of her lacks and the lacks of her life, which in turn create her needs for moral characters like Rodolphe and for more women's novels. Like Charles and like all the other bourgeois characters in Flaubert's novel, Emma's hopes for social advancement are predicated on a faith on documentary evidence, though of a past, not of a present.

We may recall again Bruno Latour's characterization of maps as providing performative and prescriptive guides for moving through the real. With the novels that Emma reads description is prescriptive and the prescriptive is descriptive; reference and sense are poles of cognition and action. Below is a well-known map of life's romantic journey for women to follow, from Madeleine de Scudéry's romance novel, *Clélie*. The map depicts a woman's prospective travels from a new love to the "dangerous sea" and the unknown lands beyond, passing through bad and good moral attributes and the large "Lake of Indifference" in the middle of her journey (figure 4.1).

Emma's life is caught within a genre war between men's documentation and women's literature, that is at times equally fictitious and real in turn, depicting science and medicine and critical and philosophical thought on the one hand and desire and passions on the other. She attempts to transcend this bourgeois sphere with actual romance in order to gain the certainty of aristocratic freedom. She uses the "map" of earlier women's romance novels to do this. Faced with hypocrisy and uncertainty in the realm of reason, Emma turns to the "certainty" of desire and sensuality, in literature and in life.

As is shown also in Jane Austen's novels, in Flaubert's novel there are pictured the two gendered sides of the literate canon in modernity: on the one

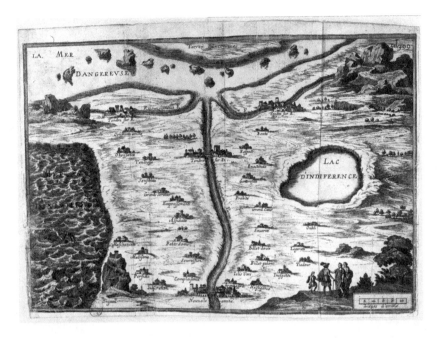

Figure 4.1
Carte de Tendre, from Madeleine de Scudéry's ten-volume 1654–61 novel *Clélie*.

hand, the "best authors" who represent "true" or "considerate" thought in science and philosophy, and on the other hand, the moral spheres of the passions in novels. Two quotes about libraries in *Madame Bovary* serve to depict the literary divide that marked off "serious" literature from romance tales. Early in *Madame Bovary*, the character of Leon first meets Emma at the home of Yonville's pharmacist. While Leon and Emma flirt with one another, discussing fiction and poetry of the emotions, the pharmacist, who imagines himself a man of educated reason and taste, awkwardly interjects about the availability of his library, which offers everything on natural, social, and moral fact, excluding of course the very type of discourse and ways of life that are being performed in front of him by Leon and Emma. With this interjection, Leon and Emma's two-and-a-half-hour flirtation comes to a dead stop. Flaubert (1904) writes,

> "Has it ever happened to you," Leon went on, "to come across some vague idea of one's own in a book, some dim image that comes back to you from afar, and as the completest expression of your own slightest sentiment?"

Documentarity and Modern Literature

"I have experienced it," she replied.

"That is why," he said, "I especially love the poets. I think verse more tender than prose, and that it moves far more easily to tears."

"Still in the long run it is tiring," continued Emma. "Now I, on the contrary, adore stories that rush breathlessly along, that frighten one. I detest commonplace heroes and moderate sentiments, such as there are in nature."

"In fact," observed the clerk, "these works, not touching the heart, miss, it seems to me, the true end of art. It is so sweet, amid all the disenchantments of life, to be able to dwell in thought upon noble characters, pure affections, and pictures of happiness. For myself, living here far from the world, this is my one distraction; but Yonville affords so few resources."

"Like Tostes, no doubt," replied Emma; "and so I always subscribed to a lending library."

"If Madame will do me the honour of making use of it," said the chemist, who had just caught the last words, "I have at her disposal a library composed of the best authors, Voltaire, Rousseau, Delille, Walter Scott, the 'Echo des Feuilletons'; and in addition, I receive various periodicals, among them the 'Fanal de Rouen' daily, having the advantage to be its correspondent for the districts of Buchy, Forges, Neufchatel, Yonville, and vicinity."

For two hours and a half they had been at table; for the servant Artemis, carelessly dragging her old list slippers over the flags, brought one plate after the other, forgot everything, and constantly left the door of the billiard-room half open, so that it beat against the wall with its hooks. (pp. 103–104)

The second quote that refers to libraries in *Madame Bovary* that I would like to point to comes from an exchange later in the book between Charles and his mother. Charles is concerned once again with Emma's rapidly shifting moods and poor health, which Charles blames on the environment or on her innate constitution. His mother, however, blames Emma's emotional turpitude on her lack of work and spending too much time reading. Flaubert (1904) writes,

Then he wrote to his mother begging her to come, and they had many long consultations together on the subject of Emma.

What should they decide? What was to be done since she rejected all medical treatment? "Do you know what your wife wants?" replied Madame Bovary senior.

"She wants to be forced to occupy herself with some manual work. If she were obliged, like so many others, to earn her living, she wouldn't have these vapors, that come to her from a lot of ideas she stuffs into her head, and from the idleness in which she lives."

"Yet she is always busy," said Charles.

"Ah! always busy at what? Reading novels, bad books, works against religion, and in which they mock at priests in speeches taken from Voltaire. But all that leads you far astray, my poor child. Anyone who has no religion always ends by turning out badly."

So, it was decided to stop Emma reading novels. The enterprise did not seem easy. The good lady undertook it. She was, when she passed through Rouen, to go herself to the lending-library and represent that Emma had discontinued her subscription. Would they not have a right to apply to the police if the librarian persisted all the same in his poisonous trade? (pp. 157–158)

Repeatedly in *Madame Bovary*, the main characters express themselves through the indirect and direct mediation of texts. The moral spheres of textual evidence originate their actions and socially and psychologically position them. Bourgeois social space, unlike the hypocrisy of aristocratic space, as demonstrated by the actions of Rodolphe, is shown in the novel as made up of rules and roles for behaviors that are not expected to be the façade of courtly or aristocratic life, but rather the core of the self and its agency in finding and securing an earned and rightful social position. This is what Emma finds so stifling and what clashes with her imaginative flights and desires. Emma's fight against the documentary nature of bourgeois social identity by means of flirtation, dialogue, and the novels that valorize such mirrors the tensions in Flaubert's own book, which is both a romance novel and itself a realist document of bourgeois life at the time and place of its composition.

Madame Bovary offers an example of novelistic realism that, in comparison to earlier literatures and scientific accounts, intensifies evidential depictions of reality (particularly at the points of expanded emotional content) and focuses upon powerful particular agents and the complex social interactions of such agents. As fiction, the book offers a descriptive and prescriptive model by which readers can understand a former time and place and still find directions for choices in their own lives because of shared class morals. Morals show that little has changed in the construction of class since Flaubert's time, despite radical changes in the technology of modes of production.

As Doty and Broussard (Broussard & Doty, 2016; Doty & Broussard, 2017) have argued, information can take fictional forms that model and guide human behavior. In this book, I have taken a broader notion of "information" than that of twentieth-century information theory: that of the document as evidence, and the still-broader notion of the philosophy of evidence as constitutive of Western metaphysics. I suggest that modern realist fiction must not only be accounted for as a counter-discourse to modern notions of documents or information grounded in an epistemology of fact, but also that we must see such literature as part of a dialectic within a logic of the representation of reality through forms of evidence (documentarity). The evidence offered through modern realist fiction is that of a model that emphasizes expanded emotional content, the power of particular agents, and complex social interactions in time and space specific horizons. Particularity and phenomenological completeness of description are intrinsic to its descriptive power.

Later literary and art traditions, which must be accounted for in modernity in regard to documentation and information, are those that emphasize not the modeling of emotions, but their appearance in aesthetic shock, and also that of the construction of reference and sense by means of formal innovation ("technique"). In shock, the machine of aesthetics is seized and used against harmonious affects, and in the second tradition, formalism and constructivism, the devices of aesthetics are used for new, nonrepresentational ends. Both these modes constitute the modern literary and artistic avant-garde of the twentieth and into the twenty-first centuries. Powerful particulars here are less depicted and more performed, becoming agents for constructing the future.

The Modern Avant-Garde

With the modern literary and aesthetic avant-garde, we see a critique of representational realism as not just a literary genre, but as a "social ontology" within a semiotic field (e.g., political discourse or online information and communication). The realist commitments in literature parallel those in documentation in the "strong" semantic techniques of classification, and they reemerge through algorithmic mediated searching and subject indexing through probabilistically calculated needs and use. As the title of the writer Carla Harryman's literary book puts it, in different strong and weak documentary manners, these are literary and nonliterary commitments to evidence and being in the metaphysical and political "mode of," which is to say, in the language of this book, in the mode of documentarity. In the theory of the avant-garde, however, performative technique interrupts realism and the dominance of semantic space in our lives with "disruptive" performances against normative, containing ("of"), frames of representation, and suggesting new forms. Realism is inverted from being the frame for seeing reality to being one of many different types of inscriptional devices, and so the empirical real is opened up again within a semantic space depicting agency, relations, and action.

The nineteenth-century French realist novel critiqued, but also intensified, documentary realism by creating characters with intense and subtle emotional range and internal mental dialogue, situated within, and responsive to, modern social conditions (not least, created by mass media). Like today's "information," it closed down the space between the reader and itself as document, inscribing the reader in the world of its own senses, leaving the reader to enact interpretive reading in the analogical application of its "content." Reference was intensified by means of complexities of sense, both in regard to the internal states of the characters and the world around them. In contrast, the modern avant-garde, particularly in the early twentieth century, pushed toward a counter-discourse of evidence rooted in sensation and in the creation of new forms out of semantic materials. "Characters" in the modern avant-garde are largely the viewers and the readers themselves, along with the artist or writer, in acts of performing, rather than representing, situational affects. Consequently, the realm of "fiction" as a mimetic representational space is replaced by empirical or "material" events. Ideological horizons represented in historical fiction are made present by performance in the avant-garde.

Dada and its presentation of the artwork as a shock effect stressed the materiality of performance against modern Kantian aesthetics' erasure of materiality in the latter's adulation of content in the harmonious form of the beautiful. With this emphasis upon materiality there also came an emphasis upon performance and the site and time-specific nature of the performance. The degree to which performance and sensation by means of the work in the artistic and literary avant-garde was seen as an end or "event" in itself or the degree to which such materiality acted as a mode of defamiliarization toward social, cultural, and political reconstruction (as in Soviet Constructivism) varied. In the performative tradition of the avant-garde, which could to be said to reach from earlier twentieth-century Dada up through later twentieth-century Fluxus and performance art, shock value and the materiality of objects, sounds, lights, and the human body were valorized in displaced settings.

Documentarity and Modern Literature

Hugo Ball's 1916 poem of nonsense sounds, "Karawane," for example, offers imagined "words" whose sounds depend upon the language they are interpreted within. Different readers apply different linguistic rules or habits upon the written script. Readers naturally attempt to make sense out of these "nonsense" inscriptions (figure 4.2).

Unlike representational aesthetic works, which in Kant's (2000) formulation create harmony or disharmony in the viewer or reader's mind by corresponding to pleasurable forms (the beautiful) or by their falling outside of the harmonious boundaries of representation (the ugly or sublime), the art and literature works of the modern avant-garde prioritize the experience of

KARAWANE
jolifanto bambla ô falli bambla
grossiga m'pfa habla horem
égiga goramen
higo bloiko russula huju
hollaka hollala
anlogo bung
blago bung
blago bung
bosso fataka
ü üü ü
schampa wulla wussa ólobo
hej tatta gôrem
eschige zunbada
wulubu ssubudu uluw ssubudu
tumba ba- umf
kusagauma
ba - umf

Figure 4.2
As we see with the text of "Karawane," the work's contestation of meaning can occur in different registers of materiality: linguistic meaning and visual meaning (and sound meaning, when read). Visual language here gives the possibility of verbal contestation and invention.

the viewer in relation to the material of the work in the face of normative expectations. They are less representational, and more presentational. This creates a self-reflection of the viewer in the face of the work, as the work is posited as a reworking of normative conceptions and values for the meaning and use of language and imagery. This tradition continued through mid-twentieth century minimalism (where the meaning of the work as an art object was contested, often within the frame of art institutions) and conceptualism (where the concept of the meaning of the work is contested by distortions or extensions of meaning and scale in site and time specific contexts [Watten, 1984]).

In conjunction with the purposeful disruption of meaning, there is also a tradition in the avant-garde of arbitrary or aleatory techniques (e.g., later in the twentieth century by poets such as Jackson MacLow or Clark Coolidge), which both isolate linguistic units and, like Ball's works, draw in the cognitive and sociocultural frames and assumptions of readers. Intentional representation by the author is minimal or absent in such works, with both sense and reference being largely created by the source text and reader response. Multiple horizons of possible sense and reference are available for the reader to draw upon, thanks to language, common experiences, and publicly mediated discourses, and the author makes use of these horizons for either empowering the reader and/or challenging these horizons in a contestation of meaning. Here, for example, is an arbitrarily chosen section from Clark Coolidge's 1974 book *The Maintains*, a book composed of aleatory chosen language drawn from a dictionary:

> such like such as
> of a whist
> a bound
> dull
> the mid eft
> lulu
> the mode
> own of own off
> partly of such tin of such
> the moo
> which which
> lably laugh
> meter it's too
> too maybe

lately too
same the marge
noun
by down which say
such way
ken ablative
sand's off
the lend the so
can which of
townly

(Coolidge, 1974)

The Maintains has one foot in performance and another in a constructivist rebuilding of meaning using social horizons (Watten, 2003). This dialogue between language and its reception in social space depends upon writing as technique, and it is upon writing technique as a critical apparatus for social construction that the avant-garde, from Russian formalism through constructivism and then through recent Language Writing (sometimes called "Language Poetry"), focuses on, and which we turn to next.

Writing as Social Contestation and Construction

Russian formalist theory and constructivist practice in the first quarter of the twentieth century gave priority in art to technique in the construction of new meaning. Constructivist techniques in art and literature, when applied in a tactical manner to social and cultural materials, can be used to build new forms for meaning.

Russian formalist theorists such as Victor Shklovsky, Yuri Tynianov, and Roman Jakobson theorized that there were poetic devices, such as metaphor, metonymy, rime, alliteration, and assonance, that worked to not only cohere rhetorical units, but to also defamiliarize normative meaning. In Russian constructivism, Russian futurism, and in allied movements across the arts, literature, and film, formal technique was seen as not only capable of defamiliarizing normative meaning, but it could also be used for generative practice in order to construct new meaning. For the formalists, these "poetic" devices were the sources of all literature. The Soviet artists and writers saw themselves as participants in creating a new society through such making or *poiesis*, using these devices for new ends. Rather than intensifying (realism) or "shocking" old meaning (Dada), and rather than simply reducing meaning to its material base of language (sound, letters, etc.), constructivism attempted to reconstruct meaning at the level of its semantic materials in order to reconstruct society. Documents are taken as the materials of art toward examining their meaning and in creating new meaning. The representational assertions of nineteenth-century realism and naturalism are rejected, and instead a new "realism" (composed of empirical semantic materials and their devices) is shown and engaged. These formalist and constructivist traditions in art and writing continued through the twentieth century and are still prevalent today. In literature, the movement of "Language Writing" (or "Language Poetry") from the 1970s through today continues the formal and constructivist traditions, and so it is to some examples from Language Writing that we will now turn.

Language Writing

Barrett Watten (2016), one of the leading writers and theorists within Language Writing, has argued that, in Language Writing, the "radical particular" is important. From Objectivism in American poetics earlier in the twentieth century and continuing through New American Poetry in the 1960s, there was a concern with the radical particular, and especially in New American Poetry, the radical particular of the self. Louis Zukofsky's works and William Carlos Williams's implementation of his famous phrase "No ideas but in things," would be examples of particularism within American Objectivism in the mid-twentieth century; Robert Creeley's concern with the self and with linguistic indexicals or demonstratives, as well as with presence and place, or the works of Charles Olson, would be examples in New American Poetry after Objectivism, with particular emphasis on the psychological self rather than linguistic objects or "particulars." And, of course, the "Beat" writers were famous for viewing writing as descriptions of individually lived experience.

For Watten, however, a poetic focus upon "radical particulars" in poetry brings with it a negation of totalizing form or argument and a leveling of ideational constellations. In Language Writing, the radical particulars are the materiality of language or the identity and agency products of that. Rather than focusing upon the empirical particular as leading to a radical departure from a poetics of concepts (as Williams or Olson did, in contrast to "academic poetry" in their eras), Language Writing has focused upon contesting both modes, contesting the sociocultural construction of social empirical entities and their concepts. Watten writes in his book *Questions of Poetics: Language Writing and Consequences* (Watten, 2016):

> At the basis of Language writing's productivity are its three most distinctive features: (1) *radical particularity*, the making of poetic form out of the serial accumulation of myriad particulars, each a differential fractal of a larger form; (2) *aesthetic negativity*, the critical distance from original contexts taken by the making of radical particulars and their contestation of larger forms of organization and subordination; and (3) formal agency, the critical alterity and interpretative openness of the work. (p. 8)

Particularly in the second chapter of *Questions of Poetics*, Watten discusses the radical particular as a linguistic unit in regard to poet Ron Silliman's notion of the "New Sentence" (as a sentence indexed to social and material fragmentation in late capitalism) and Carla Harryman's mixing of genres and her leveling of hierarchical orders through narrative (which we will soon examine). The New Sentence is but one technique by which genre assemblages such as the lyric and story are challenged, as the stress upon the assertion of the sentence, above the hierarchal or organic whole of the paragraph or poem, is prioritized. Watten writes,

> The crux of the relationship between technique and method in Language writing, between its radical particularity and open form, is its privileging of the part over the whole, particular over universal, signifier over signified—the formal "dominant" found everywhere in the work. (p. 85)

Watten's own poetic works have often marked the "missing x" of particular agencies and materials in political space by engaging the problem of what is absent in public forms of information (such as the media). His poetic work has been a continual attempt to think the "not" in both its negation and its affirmation (including through the construction of the poem itself as a social act). While this has been a problematic for the historical avant-garde generally during the twentieth century, Watten's poetry and scholarly works put to the fore the tensions inherent within social and political positioning via writing or inscription, rather than finding comfort in assumed subject positions via reifications of "the body," "self," or "experience" (as was the case in performance art, New American Poetry, and Beat poetry). In the place of propositional statements or assertions (or as we have characterized their larger form, documents) as evidence of what is, Watten offers their negation, which, in his long poem *Progress*, results in a positioning of knowledge (and the self) as everything that "is not the case" (inverting Wittgenstein's famous assertion in his *Tractatus Logico-Philosophicus* that the world is everything that *is* the case). In *Progress*, the self is not a representation, but the absent index of documentary assertions made in political and social spaces. The ironic progress (the negative dialectics) of the poem gradually reveals the self as what is left out of what is socially reported, and politically functions, as the represented world. "Progress? / To identify a body by pain / Of cultural space inscribed" (Watten, 1985). Through the form of the poem the radical particular appears not only in linguistic elements, but also in the negated self in the midst of its continual mediation (and finally, abjection and exclusion) by mass media.

Another example of Watten's critical poetic engagement with public information and media is his 1988 book *Conduit*. In *Conduit*, Watten critically engages what Michael J. Reddy (1979) termed the "conduit metaphor" for communication and information (e.g., in Warren Weaver's interpretation of Claude Shannon's information theory, where communication is assumed to be the transfer of information from one individual mind to another). With the conduit metaphor, what is left out is the site and time specificity of expression and understanding, the cultural and social forms for any meaning, and the "mind" as a cultural form or toolbox of expression. (The conduit metaphor has been a mainstay model of cognitively based information theory in information science, despite its early twentieth-century origins in folk psychology [Day, 2000].) Watten's long poem contests the conduit metaphor by supplying the invisible labor of meaning creation. Argument remains in what is left between statements, rather than organizing statements into meaningful memes of "information." Here are excerpts from the introduction to *Conduit* and from the beginning of the poem:

> The world is structured on its own displacement. "We don't believe our senses. The level of automatism we have to deal with . . ." is functionally exact. There is a continual need for new forms through which this distance might be converted into a formulation of the immediate present. The present no longer appears likely in the form of an identification; rather, assertions mark the limits that identity can only fill in. For if the world were only what it is, there would be no place for us.
>
> . . . The forms of a riddle travels through spaces and time. We question a question in order to fill in its form. Its meaning is the questioning act. If "existence" is calling itself into question, we can easily supply the answer because in that case we know; the question has become ourselves. If "existence" is the question,

writing will be perceived insofar as that is the question it asks. Here there can be no objects of thought but only an extension of the temporary that effaces any motives. Then the world is only this kind of instantaneous act? Its history falls like an oily rain. Only a rigorous avoidance will tell us anything (will tell us "it is like *that*"). Fashion models twist and turn in front of the camera as the shutter clicks. The public reads Sartre on busses. We make something out of what's missing by filling in the blanks, giving our meaning to what has been negated. Such are the limits of art.

The world is everything that is *not* the case.

I
An arrival in history only coincides with defeats.

The invisible body is a mirror of containers.
Perfected, a chain of commands speaks.

Every road ends in an object. Unfolding, a
world of parts in a display of same.

> A sign revealed in buses for the driver or deportee.

> Each utterance is unique only in a theory that specifies a point in time and space.

Then all the pawns fall.

While a model faces perception. Behind every
Survival is a whole totality in words.

> I.e., parentheses it is a text. To answer a question in runic workings of quotations behind which passion slips.

> And left, imagining a sum: an *I* inverted to an *it*.
> At the same time as an impostor.

(Watten, 1988, pp. 151–157)

Here we do have aesthetics ("feelings")—not in the Kantian sense of a relation of mind to the harmony or disharmony of an artwork or to the natural world, but rather as abjections and negations of the self within and through socially and politically generated normative representations of reality. The response to this by utilizing art technique is to fill in those missing holes of where one exists and what can be done. As Walter Benjamin suggested, if the news was meant to bring ordinary people closer to events in a world of their experience that they could act in, it utterly fails. It is not meant to do this, however, but rather to distance people from their inclusion by presenting the world "as it really is"—that is, without them or their participation.[4] Abjection, not inclusion, is the goal of the mass media. "This is what the world is, and as such you don't belong in it," the objectivity of the news tells the reader or the viewer. The suggestion is always that one is not enough for the world "as it is." The subject can never measure up to the problems and needs of the "objectively" reported world, and so she or he can only participate by commenting on it or voting on "its" issues. The poetic work responds to this with techniques for reasserting the "missing x" into the frame of the real.

We encounter a different strategy of techniques than Watten's in the narrative works of Carla Harryman. In Harryman's works, narrative is used, but not in the form of traditional modern realism (i.e., typically, with a story composed of beginning, middle, and end, causal effects between characters' actions and results, moral outcomes, and characters starting from or created into social types). Such literary devices for representational framing aren't discounted in her work, but rather they are taken as materials for constructing narrative, rather than structuring such. In her works, narrative is a stream of relations, as commented on in this short fragment from Harryman's *Gardener of Stars* (2001):

> building a world of roads
> connecting everything to everything else
>
> (p. 104)

We see in Harryman's works a critique of universalization and possession as the products of the assertion of categories and transcendental essences. (The phrase "in the mode of," the title of her 1992 book, could be a synonym for what I've called "strong documentarity.") In her book *Adorno's Noise* (2008), Harryman writes,

> For me writing often involves a necessary dissimulation, dismantling, undoing, refusing, renegotiating, criticizing, and diminishing of the weight and values of symbols. Or I could say encounters with powerful symbols, including the "phallus," "the flag," "the canonical," "the famous," "the family," "the good movie," "the poet," "the argument," "the essay," "the binary," "the war," "the government," "the mall," "the museum," "the philosopher," lead to questions about what is presumed to be held in common. But what if instead, my tactics were to place symbols under a kind of magnification? Such that they expanded to the

point of disintegration, thus no longer identifiable with things but diffused in a total environment—as if particles in the air we breathe? (p. 173)

For Harryman, narrative can pressure, rather than create, representations, to the point of exploding their meaning. By "connecting everything to everything else," argument, genre, plot, and character, for example (i.e., rhetorical devices for representation) become revealed as devices or tools. Harryman's technique tends to connect particulars with one another. And narrative is built up at the level of sentences, rather than forcing sentences (and the meaning generated through them) to work from ideational categories, from top-down argumentative structures (via topic and thesis sentences), and in literature, from genre assumptions.

The following section from "The Male" in Harryman's 1989 book *Animal Instincts* shows her challenging "realist" compositional methods and their semantic and pragmatic outcomes:

WOULD YOU PREFER the examples? The pancakes? Or the words?

Oh, I have been used as an example so many times, said the Male. I think I . . . Do I? Do I think? said the Male.

Pancakes are good, I reminded him.

If, said the Male, I say anything, I reveal something of myself, my stupidity, or arrogance, or inability to make selections. I can't speak . . .

If you could only make a choice, I could say, for example, well the Male prefers pancakes, and that must mean something. Words pain the Male, I could say. And then I would attempt to apply that information as an example. Everybody would be able to make sense out of the expression *the male's pancakes*. When in the galleries, I could point to the portrait of an ancestor and say "the male's pancakes," and everyone would laugh from the pleasure that words and things can so transform each other they make the most sense when used in tandem.

. . . The Male by nature prosaic, moving from one place to the next in an unrhapsodic way, thinking hard perhaps, but communicating little, allowing his motions to speak for him, so that he was followed by a trail of his own making? Would others follow this trail, each having their own experience of it, each wondering what it was like for anyone else to have been there? (For instance, what it was like for Orphan Annie? The cranky-looking filling station out the window? The hoses on the pumps having lost their resilience? The attendant limp as grease? The comic-strip reader in a sunlit, airy place?) Life is like a book, any book, even technical manuals.

On the other hand, there is the body, a form, and who knows what goes on in the Male's mind? The Male would exhibit a deep, ponderous blank. And yet, *I* do not have a verse in any of my thoughts. Is a landowner a landowner *all* the time? The landowner would either say "yes" or "no, I'm just a person."

> I am just a person, I said to the Male, but you are not just a male. I don't know why I chose to present myself in this way to the creature. (1989, 12–13)

Similarly, in the text below, from her book *In the Mode Of*, Harryman critically engages the proposition "of" and its grammar of possession and containment:

> I like to think about prepositions as possessing character, narrative or metaphorical qualities. I think of all prepositions as being compromised, as social and autonomous, visible and barely there, subordinate and subordinating. Of is the most subordinating and aggressive pronoun. It is also the enforcer of social assumptions. Of would be a difficult person to like.
>
> Needless to say, the difference between having a vision of what one is of and not having a vision or imagination of it is significant.
>
> Or one body becomes part of another body. A body is subsumed by a body. Hierarchies and imbalances are affirmed, created, and subtly invoked. The distinction between one thing and another, one person and another, is leveled and consumed. (The daughter of Bob). Confusion, overlapping, boundary erasure. A conduit of aggression and destruction: the destruction of the people or the celebration of the people may represent two sides of the same /invisible agency. If OF were a mythological character it would be the god of illusion and instability. (1992, pp. 29–30)

In Harryman's works, literature and art contest what is seen as political and psychological containment—"power," in the containing or repressive sense of this term in English. In the avant-garde, the inversion of material form (e.g., language) and ideational content presents a "materialist" insertion into the supposed transparency of language as representation.

The critical turn of this in relation to documentation and information is that the distance between signs and what is represented—between "poetics" and "knowledge"—appears as technical and tactical. Institutional knowledge and personal knowledge are not transparent through language, but rather, they are constructed, and sometimes as representations. In the avant-garde tradition of art and literature, of *poiesis*, the construction of meaning is pushed to the forefront of any assertion. And conversely, assumed representational "information" is defamiliarized, "evidence" is questioned.

In modern literature, and more generally, in art, there are thus two very different "realistic" traditions: in most nineteenth- and twentieth-century realistic novels and smaller works of modern and contemporary fiction, realism is the extension of representation to the furthest recesses of depicted sense. Conversely, in the avant-garde tradition, the devices for representation are made available for sense and knowledge production, rather than acting as a frame for such.

From Poetics to a Critique of Information and Knowledge

Before we leave this chapter, I would like to further note, in wake of my analyses of Watten's and Harryman's works, that the tradition of critique of representation in the avant-garde has not, of course, remained bifurcated between art and literary "practice" on the one hand, and philosophical or theoretical "critique" on the other. Both Watten and Harryman have works in both domains and their works cross these domains. And in the works of artist, poet, and academic theorist, Johanna Drucker, for example, they have crossed these two traditional domains in a critique of textual, but also quantitative and visual, information forms. Drucker's works critically engage both bibliographic and data information at many levels of construction, including their technical construction in digital design.

Drucker's body of work is unique and amazingly broad in this respect; it seems informed by her early work in Language Poetry, her work as an illustrator and artist, and then her scholarship in bibliography, digital humanities, and information visualization. Her work in digital humanities and visual representation emphasize that the representative frames for information should not be taken as natural, but rather as arguments, and as such need to be revealed and critically discussed. For Drucker, the humanities bring critical and interpretative tools to information, including to the tools and assumptions of the quantitative digital humanities. Drucker (2009) writes:

> The digital humanities community has been concerned with the creation of *digital tools* in humanities contexts. The emphasis in speculative computing is instead the production of *humanities tools* in digital contexts. We, however, [i.e, Drucker and her colleagues at the University of Virginia in the first decade of the twenty-first century] are far less concerned with making devices to do things—sort, organize, list, order, number, compare—than with creating ways to expose any form of expression (book, work, text, image, scholarly debate, bibliographical research, description, or paraphrase) as an act of interpretation (and any interpretative act as a subjective deformance [*sic*]). (pp. 25–26)

These qualities are particularly important for quantitative information, where the *a priori* and the aesthetically constructed elements of argument can be most hidden. For Drucker, the representational tools of visual information should be revealed and subject to critique as devices and arguments. Drucker writes in her book *Graphesis: Visual Forms of Knowledge Production*:

> The basic categories of supposedly quantitative information, the fundamental parameters of chart production, are already interpreted expressions. But they do not present themselves as categories of interpretation, riven with ambiguity and uncertainty, because of the *representational* force of the visualization as a "picture" of "data." For instance, the assumption that gender is a binary category, stable across all cultural and national communities, is an assertion, an argument. Gendered identity defined in binary terms is not a self-evident fact, no matter how often Olympic committees come up against the need for a single rigid genital criterion on which to determine difference. By recognizing the always interpreted character of data we have shifted from data to capta [sic], acknowledging the constructedness of the categories according to the uses and expectations for which they are put. Nations, genders, populations, and time spans are not self-evident, stable entities that exist a priori. They are each subject to qualifications and reservations that bear directly on and arise for the reality of lived experience. The presentation of the comparison in the original formulation grotesquely distorts the complexity, but also the basic ambiguity, of the phenomenon under investigation (nations, genders, populations). (Drucker, 2014, p. 129)

The academic works of Drucker unmask the technical and conceptual devices that go into the construction of information and information infrastructures. Her works uniquely engage a critique of information as representation through a broad range of academic and artistic modes across a lifetime of practice. They argue for the dual use, and at times the merging, of critical theory and artistic practice in a critique of information and knowledge representation, and for a material practice of revealing the devices of representation in art, information, and scholarship.

As discussed in the introduction, after looking at documentary in the beginning of this book from the aspect of *a priori* categories, we end this current chapter from a "near-zero" point of representation and a poetics of critique and constructivism. In the next chapters we will examine documentary more from the aspect of performed expressions and their affordances and *a posteriori* inscriptional registers. Beginning this turn, in the next chapter, we briefly look at two subgenres of literature (jokes and fables), as well as at the psychoanalytically derived theory of trauma, in order to see expression performatively appearing from implicit or repressed premises through literary-social devices.

5 Displaced Reference for Information: Jokes, Trauma, and Fables

Before turning to powerful particulars as ontologically expressive agents, I would like to look briefly at powerfully particular expressions within signification, conceived of as expressions of repressed or potential meaning—meaning that appears from below the bar of overt signification (thus being sublime, potential, or latent powers). In this chapter, I will describe three cases where evidence appears through literary and narrative devices in performative social space: jokes, fables, and the psychoanalytic theory of trauma (which is based on a notion of recursive time).

In each of these cases, literary-social genres play an important role in attributing meaning to an expression. Rather than the universal subsuming the particular in essentialist assertions, it is the universal that appears in the particular, though it is the assignment of statements to genres or normative contexts in the background that afford this. So, for example, unless we understand a set of statements as a joke, it is difficult to perform the surprise, or "witty," elements of jokes; unless we understand events as traumatic events (following the psychoanalytic theory of trauma as deferred manifestation), then we simply have negative affective events in the past and/or aberrant behavior in the present; and unless we have a "once-upon-a-time" framework or some similar "as-if" structure common in fairy tales, then what we have is an imaginative story that is past. However, as I will argue, each of these performances also depends on the actualization of meaning in social space. (So, e.g., jokes are not "gotten" unless the listener is familiar with not only the cultural forms, but also the social functions, of a speech act.) They function not only as literary, but also as more broad social genres and devices.

Jokes

Jokes are usually not considered to be knowledge and/or information phenomena.[1] Their status as evidence is even derided as often prejudicial or in bad taste. This, however, is also partly why jokes usually belong in the realm of comedy rather than tragedy, for jokes are a type of expression that resolve in absurd figures of speech or reference.

Jokes often occur as descriptions of states of affairs. But these states of affairs exist in an unexpressed premise that appears in the process of the joke; jokes function like enthymemes. They are thus a special mixture of performance and representation combined. As noted in Sigmund Freud's famous 1905 book on jokes, *Jokes and Their Relation to the Unconscious* (Freud, 1989), jokes may be ways of asserting statements about people or events or more abstract states of affairs (e.g., stereotyped national, cultural, or gender dispositions) without speaking "directly" on the matter. In this, they also mark themselves as exaggerations, though with the caveat of all exaggerations, of course, that these exaggerations also claim to be pointing to states of affairs.

Here, I am interested in investigating how evidence is asserted through the ironic and performative nature of grammar in jokes. I am particularly interested in those jokes in which an "unconscious" or nonsyntactically normative meaning to a word or phrase appears as a surprise to the listener, and so may cause laughter by the performative perversion of normative meaning by the joke. In jokes, a perverse function of grammar plays a role in creating meaning, either through creating an odd meaning (via an ambivalent word or phrase) or by an odd reference. Like many of Freud's examples of jokes, I will concentrate on jokes characterized by wordplay.

Freud's *Jokes and Their Relation to the Unconscious* is notable not only for Freud's own lightheartedness and wit, but also because the third part of the book, that of a theoretical account of jokes, demonstrates a linguistic understanding of the psychoanalytic notion of the unconscious. Specifically, jokes are presented in terms of semantic substitutions in scripts, linked by ambivalent terms and background premises. If, as Jacques Lacan famously said, the unconscious is structured like a language, then it is the genre of jokes that gives the impression that there exists such a thing as a subliminal faculty of meaning—"the unconscious." The assumption of all three genres in this chapter—that there are "repressed" entities that are indirectly being referred to and retrieved—is made possible by techniques of grammar.

The sixth chapter of Freud's work *Jokes and Their Relation to the Unconscious* sets up an analogy between jokes and dreams in terms of their basic mechanisms. The two basic mechanisms for jokes are those that, in his famous work published in 1900, *The Interpretation of Dreams*, Freud (1980) had set out as the two basic mechanisms of the unconscious, as well, which he saw as evident in dreams: the *condensation of signification* from two or more terms and the *displacement of signification* from one signified to another, resulting in an indirect manifest representation of the reworked latent meaning of the dream so as to avoid both the "censorship" of the preconscious and the awakening of the sleeper. (Or, in the case of jokes, avoiding making direct embarrassing assertions that would "awaken" the listener to the speaker's prejudices, for example.)

Freud's comparison between dreams and jokes provide us with two very important understandings of jokes as meaningful events. First, the primary device for "unconscious" meaning to arise in a joke is a linguistic switch that links two or more scripts. Second, earlier in his book, in the third chapter, Freud distinguishes between non-tendentious and tendentious jokes: jokes that have as their primary purpose merely the release of a social or grammatical tension through the play of the joke itself (in Freud's terms, a release of cathartic energy) and those that have some more overriding purpose (e.g., to insult the listener or to malign some social group). In his comparison of jokes with dreams, inevitably the comparison leads to emphasizing tendentious jokes, since the very purpose of dreams for Freud is to rework difficult daily issues so that they may be made acceptable (to social taste, in the case of jokes).

Focusing upon jokes as linguistic play, I will examine some old jokes and ask how they can be analyzed as linguistic condensation and displacement, so that a nonnormative or "unconscious" utterance is released by the joke. (The "information" of the joke may be taken as this release of a "latent" into "manifest" or expressive content, and it may also be taken as the performative utterance itself—say, as socially indexing the speaker as a wit or comic.)

Our first joke goes as follows:

> A man walks into a restaurant with a lobster and is seated. He says to the waiter, "I'd like to have my lobster for dinner." The waiter responds, "Would you like to have the lobster boiled or baked?" And the man responds, "Oh no, just bring him a steak."

This joke functions through the ambiguity of the term "to have" in the context of the joke's narrative. The normative or "major" meaning of "I'd like to have my lobster for dinner" would be that of eating the lobster. The "minor" meaning of "to have" in this context (i.e., to have as a dinner guest) as the dominant meaning, however, is revealed at the joke's end, namely, when the man asks the waiter to bring the lobster a steak to eat.

Like almost all linguistic jokes, the linguistic play revolves around multiple social norms for the use of its words (or as Wittgenstein called such language in use, "grammars"). If someone were not to understand the joke, then we would have to explain it by reference to the different grammars of "to have" as used in a restaurant context, and perhaps also, by an explanation of social norms in modern restaurant cultures. The element of surprise and amusement that we experience with this joke is a function of the minor script or grammar turning around the "joke device" of "to have," which has ambivalent senses and referents that are called to our attention by the joke.

Let us take another example, this time of a type of joke called insults or put-downs. Our example comes from a collection compiled by Louis A. Safian entitled *The Giant Book of Insults* (Safian, 1981). The insult is carried out by two successive clauses, the second which strengthens the first clause, but this time through a double entendre:

> He's a confirmed liar—nothing he says is ever confirmed.

Here, "confirmed" plays two roles: that of stating identity via consistent intentional behavior by a person (as in English, "to be a confirmed bachelor") and "confirmed" in a second sense, the confirmation of statements by evidence. The play upon "confirmed" is so complex in this insult—since the second clause confirms the truth of the first clause, as well as plays on the meaning of the first's use of "confirmed"—that we may be left admiring the sophistication of the insult's rhetoric as much as the wit of its statement.

The evidential and indexical difference between insults and non-insult jokes is worth noting. Ultimately, the latter rely on shifts in socially premised situations that "switch" via ambiguity in language between two or more grammatical scripts—for example, the situation of having a lobster as a dinner guest rather than as the meal. Insults may, too, occur through such rhetorical "switches," but their ultimate indexical placement for evidence is upon the person making the insult, even as the insult is directed upon someone else. An insult is made by person x upon person y. For this reason, insults can also arise from subjective cynicism and/or feelings of superiority toward others. Ordinarily, jokes often have at least a surface neutrality that insults don't have. Their wit is meant to put them at a distance from bold assertions of fact or even judgment. Insults do the opposite. Both, however, arise by the exaggeration of imagined or real states of affairs. Insults are made to intentionally hurt another person. A joke, not necessarily so.

Recalling our previous discussion of literature, we may note here the intersection of rhetoric and psychology in joking in the developmental line leading from joking to insults to outright hatred, using the case of the French author Louis-Ferdinand Céline's writings between the two world wars. From a rhetorical perspective, the rather-amusing skepticism and cynicism (which can be seen as a form of joking involving irony) regarding the sadness of everyday life, militarism, and colonial capitalism in his 1932 novel, *Journey to the End of the Night* (2006) turns into the very mentally disturbed, vehement anti-Semitism of his four political pamphlets from 1937 to 1941, beginning with *Bagatelles pour un massacre*. In this transition, the index for "truth" becomes more and more located in the speaker's locutions, even as such locutions recirculate well-known prejudices at the time and place of their expression. Joking turns evil as it valorizes the "minor premise" of a grammar rooted in not only exaggeration, but also hatred. The insecure self is "secured" by social prejudice and rhetorical elisions. Everyone is vile and idiots so that the self can be knowledgeable. What start off as exaggerations brought about by the author's hysterical relations to the equally mad events around him eventually culminate in paranoid exclamations and their anti-Semitic materials, all utilizing streamed fragments taking the form of a rant.

If there were a rhetorical predecessor for political paranoia and hate speech in social network media today, such as on Twitter and from the highest levels of US politics, we would certainly have to look to Céline's writings and their rhetorical forms and their psychological and political functions. Across his oeuvre, we can see the rhetorical method of joking progressing to narcissism, hatred, and ignorance, asserting the most vulgar prejudices and hatred as fact in fragmentary and suggestive rhetorical forms. The purpose is not to avoid social bad taste, but to assert bad taste as the basis for knowledge and political reference, to enact a rhetoric of resentment and hatred as a state function, to replace institutional knowledge by "information"—prejudice—that everyone already "knows." It builds a security state out of the mouths of insecure egos.

Sadly, these rhetorical strategies are as effective now in the mass media and political spheres as they were before the Second World War. As I write this, the US president takes such "joking" as the mainstay for knowledge and information, and so other such politicians in the United States and in other countries follow suit at his success. The "minor" premises of prejudice are asserted as the triumph of folk wisdom over knowledge institutions.

Trauma

If jokes involve a semantic play between major and minor premises and manifest and latent content, then trauma, by contrast, relies upon the reworking of semantic manifest content by a temporally earlier latent experience.[2]

Seen psychoanalytically, trauma is an experience that creates the language through which it is expressed, a language that is also formally "traumatic" in its rhetorical overdeterminations of meaning through unexpected schisms, breaks, and double references. The psychoanalytic concept of trauma, like the literary genres we have studied in the previous chapters and this one, is missing from documentation and information studies, at least in part due to the dominance of a social definition of information that connotes temporal immediacy in epistemic or sensory presence.

In contrast, Freud's work is foundational in understanding the role of trauma as something informative through non-immediate after-affects (*Nachträglichkeit*),[3] as a primary moment of the deferred informing of the real for the subject. According to the psychoanalytic theory of trauma, what experience gives us is evidence of what has happened in the past, coloring the empirical present.

Often, according to psychoanalytic theory, the traumatized subject seeks the proper container for trauma, which by definition exceeds the subject's capacity to contain the traumatic event, and so the proper container is never found by the subject but instead is continuously displaced by the subject's past experience now projected onto objects, people, and events. Evidence of the subject's trauma is always found by the subject, but in partial and displaced forms. The original index of trauma is marked in the subject's form of speech and traced in the displaced trails and trials of the subject's perceptions. Trauma informs the subjects and everyone else around them by continual displacements and irrational fixations of the subject. It announces itself as a referent not directly, but indirectly, by objects and events, and indeed by the subject him or herself, transformed into a symbol of traumatic expression. Trauma has a tragic structure, because the subject represents him or herself as driven by a historically formed will before and beyond them, which colors the future in terms of the original events.

One of the classic sources for this understanding of trauma can be found in Freud's 1920 discussion of war trauma in *Beyond the Pleasure Principle* (Freud, 1961). In that book, Freud asks why it is that if dreams function as wish fulfillments do soldiers then reenact war traumas in their dreams. Freud's explanation is that the compulsion to repeat the trauma—in the literal forms of a dream or in the displaced forms of waking life—constitutes an attempt to master events that threaten or have threated the subject.

According to all that follows from Freud's notion of deferred action (*Nachträglichkeit*), the status of documentary evidence here lies in the way that we can read back the trauma from the expressions, while also realizing that this reading back constitutes a reconstruction from the present, as well. Documentary evidence thus lies in an act of interpretation upon an absent or latent evidence that is nonetheless present in the manifest forms of expression, which itself colors the original event in an attempt to master it in the present. Only by the illogic of the subject's actions, by their overdetermination of signification upon the real, and, in short, by all the symptoms of neurosis applied to the subject's present actions, can we assert the affective force of a traumatic event and its literally informing inscription of the subject.

Beyond psychoanalytic interpretations proper, clinical practice continues to seek traumatic causes in neurotic actions according to theories of deferred affect. In the therapeutic setting, verbal discourse has been emphasized with adults, but in the case of children, objects have been suggested as being useful as vehicles for eliciting the expression and even the causes of trauma, and reenactment is seen as useful for both children and adults for therapeutic goals. Recently, affectively responsive and recording robots, such as the robotic toy dog Therabot (Duckworth et al., 2015), have been added to the mix of tools for trauma reenactment. Such objects are claimed to allow the subject to express what he or she cannot express to another person or even to him or herself. Through the intermediary of an animal or, perhaps, even a toy animal, the subject is said to "gain a voice" to express (what is assumed to be) a real experience that is otherwise lost other than in displaced symptoms. The animal is thought to speak what the subject cannot say, and it is thought to say what the subject cannot bear being heard.

We must bear in mind, however, that traumatic expression conceived in this way is only possible if we accept deferred action as a psychological mechanism, with or without the formal psychoanalytical discourse from which the concept, and thus a notion of trauma as deferred action, historically have emerged. The theoretical construct of the past as something continuous, much less as something returnable, is an explanation that depends on narrative, historiographical, conventions. This isn't to say that trauma doesn't exist, but rather that its explanation, and possibly its experience, too, as after-affect, requires that psychological temporality be understood as continuous and durational and as explainable in terms of narrative conventions of temporal return from a durational past. The evidentiary structure of trauma can be found in ancient drama and philosophy, for example in Sophocles's *Oedipus* trilogy or Plato's theory of remembrance. Time must be seen as continuous in order for component parts to be retrieved from its series.

Fables

In the case of the psychological account of trauma, evidence is seen as simultaneously emerging from its sites of displacement and manifest absence.

As we have seen in this chapter, "information," or evidence upon which we act, is not always seen as overt in its presence. In some genres of literature and psychology, information and evidence are seen as emergent through indirect symbolic or allegorical forms, and in the modern period by psychological displacements and overdetermined signs.

In fairy tales, folktales, and fables, allegory plays the role of creating hypothetical, "as-if" structures to events being narrated. These are representational stories, like we earlier analyzed in the case of the realist novel, though their "as-if" structures are obviously fantastic and their modeling is thought to be largely "unconscious." Various versions of the tale of "Little Red Riding Hood" have been recorded, all which seem to take the theme of a warning against deception. Such tales exist precisely in order to warn children of dangers in the world and suggest mechanisms for their safety. The imaginative tales and their allegorical, fantastic, and "once-upon-a-time" narrative structures give a form of information whose interpretation is meant to fluctuate in regard to empirical referents. Rather than being realist narratives, they are allegories whose expression is double: both manifest in the tale as it is told and latent in its actualization by the child in their encounter with the real.

If trauma is marked by the extension of the "as" into daily life, then fairy tales and other such stories of literature are meant to restore the "if" into present and future events. Rather than have referents that drive the subject, they are seen as giving an availability of an explanation to the subject in case they are needed or are specifically "triggered" by events. They are documents that give formal evidence in their "as-if" or allegorical structure of the difference that such stories possess between their materials and their actualization or "meaningfulness" in the real. They are documents because they give evidence *of* something, *as* something, but *as if* it were that thing, and their status as literature is verified by the literariness by which they show this activity in their very form.

Jacques Derrida, in his last seminar "The Beast and the Sovereign" (Derrida & Bennington, 2009), engages the question of what happens when information (say, political information) takes the form of fables, becomes "fabulous," becomes *as* without the rhetorical devices of the "if" with fables or the manifestation of formal and performative literariness in literary presentations of fact:

> What would happen if, for example, political discourse, or even the political action welded to it and indissociably from it, were constituted or even instituted by something fabular, by that sort of narrative simulacrum, the convention of some *as if*, by that fictive modality of "storytelling" that is called fabulous or fabular, which supposes giving to be known where one does not know, fraudulently affecting or showing off the making-known, and which administers, right in the work of the *hors-d'oeuvre* of some narrative, a moral lesson, a "moral"? ... Well, given this, the fabulous deployment of information, of the teletechnologies of information of the media today, is perhaps only spreading the empire of the fable. What has been happening on big and small television channels for a few months now, but in particular in time of war, for example over the last few

months, attests to this becoming-fabulous of political action and discourse, be it described as military or civil, warlike or terroristic. A certain effectivity, a certain efficacy, including the irreversible actuality of death, are not excluded from this affabulation. Death and suffering, which are not fabular, are yet carried off and inscribed in this affabulatory score. (pp. 35–36)

Like in our analysis of jokes, we see here Derrida discussing the politically efficacious use of genres of evidence based on performances of sublime content. Political choices are mostly made by "unconscious" affective premises and politicians' appeals to such through enthymemic rhetorical devices. (Today, for example, through social media memes.) Here, the problem is that of applying genres that have hypothetical, "as-if" rhetorical forms, to situations where the applications are not allegorical, but real, not hypothetical, but causal. This transformation of social space into being a fable both destroys the specificity of the fable (not least as a children's genre), and also transforms the real into a real of constant crisis. The warning of the fable becomes the reality of a politics of not only crisis, but also make-believe.

As "once upon a time," as an "as-if" narrative, the fable is a guide and a lesson, and its indexical location lies between its being told and experienced. It arises from an oral tradition, not a written one. Its inscription remains in orality even when it is written. And it remains oral when it is read to a child. It remains a potentiality to appear, not a possibility that logically unfolds. Like realist fiction, it is a model, but a model that instructs through allegory.

A reality that is a fable is a reality of crisis. Is it a present that informs us as a warning allegory, whose evidence lies in a literary form.

Informational Fragments

In each of the genres that we have examined in this chapter, of jokes, traumas, and fables, what we see are fragments acting as evidence of experience. They are organized so as to produce or create surprise or lessons about experiences still to come. They suggest, they model. Indexical signs appear out of grammatical ambivalence and reversals, manifestations of latent content in time and latent content subject to time, and hypothetical tales that result in warnings and lessons about future experiences.

These genres have analogues in new media, which also bridges literary forms and experience. Fragmentary documents, such as on Twitter, make high use of contextual situations and immediate rhetorical content within them. The massively increased speed of transmitting documents, largely as small communicative rather than argumentative "fragments" through social media and the Internet, create shock waves upon waves of news fragments, which jar the temporality of everyday attempts to organize psychological and phenomenological consistency and a "private life," as we used to say. This state of "total mobilization," as Maurizio Ferraris (2013) has wittily put it, has given us a means for composing our subjectivity as the recomposition of "objective" documentary fragments according to "unseen" grammars of ideology and politics that provide and organize for us "our" world.

The role of the fragment and its ordering as sense by technical systems (e.g., social network algorithm and machine learning) in interaction with ideological horizons and popular needs continues the sociotechnical problematic of writing, now in a highly sped-up communicative and streaming archival setting (Ferraris, 2013). As a documentary fragment, its representational fragility becomes more acute, subject to extreme context variability and vagaries of rhetorical manipulation over time and audience spaces. In this, references—even what were thought of as facts—become more subject to communicative flow over time, spun like pebbles by the waters of different senses of the self and the world, mediated by political economy. On the one hand, the documentary fragment is simply the sped up and manifest showing of the communicative trails that have led to the modern conception of "information." On the other hand, as such, it has also broken away from some of the traditional institutional and genre bonds that in modernity have separated knowledge from information (such as peer review in scholarly communication and academic libraries as trusted collections of reviewed works).

In early modernity, documentary knowledge institutionally, pedagogically, and through method, separated from ordinary communicative information through the establishment of rigorous methods and institutions for scholarship and science. Today, as in the 1920s and 1930s in Europe, disruptive technologies and popular discontent with old-knowledge institutions have allowed "information" of all sorts and forms to constitute public forms of "knowledge" for ordinary people and, even at times, for scholars. Unsurprisingly, prejudice and false facts are part of this breadth of information. Later in this book, we will return to these themes.

In the next chapter, however, I would like to shift from signs as primary mediators for powers of expression, to a view of ontological particulars as agents for their own powers that lead to rights of expression. In contrast to Suzanne Briet's (1951, 2006) example of a newly discovered animal being indexed by its capture and classification, we will turn toward a consideration of the particular animal entity in terms of its own dispositional powers that can manifest as contributors to self-indexing in both worlds of other entities and worlds of signs. In the next chapter we will be interested in the social rights given to ontological "powerful particulars" (to borrow Rom Harré's phrase) to be, at least to some degree, self-evidential in science, ethics, and law.

6 Rights of Expression

In this chapter, I investigate powerful particulars as a mode of self-evidence and information, via expression within a discourse of rights. This discourse stretches from the rights of human agents to the rights of natural agents. I will perform this investigation through a discussion of dispositional "agency" rights, briefly discussing rights of information and then rights of truth before moving on to the newest extensions of rights discourse, those of animal rights and finally "rights of nature." Rights discourse is one of the older forms of viewing an entity from the viewpoint of its own singular powers or dispositions, modulated and recursively formed through "contextual" affordances.

Modern Human Rights Related to Information

Agency rights, taking human powerful particulars as their core, date from the Western Enlightenment. They are canonically expressed in political documents such as the Declaration of the Rights of Man and Citizen in France (1789), the US Bill of Rights (1791), the "unalienable rights" of life, liberty, and the pursuit of happiness in the US Constitution (1776), and as conditions for inquiry on religious matters (and by implication, on matters of scholarly public expression more generally), in Kant's 1784 essay *An Answer to the Question: "What Is Enlightenment?"* (Kant, 2013).

The great philosophical precedent for modern rights discourse is John Locke's political philosophy, where natural law and natural rights play a central role. Natural rights in this period and later assert personal agency powers of action and expression by the nature of human being, namely, in the "unalienable" condition of human free will. (In contrast with agency rights, obligation rights, in the sense of socially assigned duties, long preceded Locke's formulation.) Natural rights, understood as these innate agency rights of individual human beings, are the foundational concept for the twentieth-century notion of human rights.

After World War II, rights of information access were added to those of expression, forming the twin pillars of modern human rights and "intellectual freedom." As Fonseca and Mathiesen (Fonseca, 1999; Mathiesen, 2015) point out, information access as a human right is proclaimed in the United Nations' Universal Declaration of Human Rights (1948), article 19:

> Everyone has the right to freedom of opinion and expression; this right includes freedom to hold opinions without interference and to seek, receive and impart information and ideas through any media and regardless of frontier.

Perhaps one of the most famous information access laws is that of the US Freedom of Information Act (FOIA), which was signed into law in 1966. FOIA, however, also shows the limits commonly placed by governments on information access to government records and data. FOIA exemptions include national defense information, personnel matters in government, trade secrets and geological data, financial institution supervision, information related to ongoing criminal investigations, and FOIA requests that might violate personal privacy.

Information access is also commonly forbidden in the case of some personal information. What constitutes personal or "private" information varies widely as more and more information is mediated online and by corporate entities. Some personal information records in the United States, such as medical records, are limited by legal statutes such as the Health Insurance Portability and Accountability Act (1966).

Information access rights are usually seen as what I would call "direct rights," in the same sense that expressive rights are seen in this manner. One makes a request and, even if mediated, the information—for example, the document—is retrieved or delivered. The referent, or subject of information, is taken as a known or possibly known entity represented in the document. An example of an "indirect right" of information would be the right to truth.

Right to Truth

A somewhat-different relation to the information of documents, however, occurs with the international law principle held in many countries known as the "right to truth." The principle of the right to truth, though known in Anglo-American law literature, is practically invisible in the literature of Anglo-American information science or library science (one notable exception being Bishop's discussion [Bishop, 2012]). Such is not the case, however, in the Latin American archive and library and information science literatures and in the politics of Latin American countries and South Africa, where the right to truth emerged following truth commissions set up in the aftermath of military regimes and, in the case of South Africa, the regime of apartheid.

The notion of a right to truth in this context is the right of the family of the victims of political repression, as well at times as the national society as a whole, to have an account of what happened to particular people or peoples during times of political troubles. Beyond being personal information for the family of a victim, such information has uses for criminal prosecution, for national reconciliation, for historical accountability, and for the establishment of peace (Naqvi, 2006).

As both Naqvi (2006) and Bishop (2012) describe, though the right to truth presupposes information and an archive, it also requires the prerequisite of rights to information. However, in the case of the right to truth, information is not assumed to be only available in inscribed documentary forms, but rather it may occur in oral testimony, and the right to truth also involves broader and deeper understandings of information, toward revealing the truth of an event.

Also, in distinction to a right for information, the notion of a "right to truth" is indexed not only to the person making the request or even to the information or documentation per se, but rather to the object or subject of an inquiry about what happened to a person during an historical event. The right to truth not only belongs to those requesting the information, but also to the person or persons missing within the historical event in question. Their life—or rather the loss or absence of such—is the foundation for this right. This right can be enacted through a specific, living, person (a relative, for example) or in the name of the collective rights of a society to know the truth about a person or event for the purposes of justice or reconciliation (Naqvi, 2006).

The "right to truth" is also related to a certain notion of truth: the truth of events, rather than the truth of evidence, per se. While the first may be thought to depend upon the second, there are cases when this doesn't actually happen. For example, Medina and Wiener (2016) studied documentary errors in the forensic identification of victims during Augusto Pinochet's dictatorship in Chile, where historical identifications were based on what later turned out to be incorrect technical analyses of forensic documents. Though the documentary evidence was incorrect, the narrative provided a type of therapeutic truth for families.

One possible conclusion we may derive from their study is that though therapeutic truth in such circumstances is indexed to documentary evidence, it is not necessary for the latter to be accurate in order for the former to occur. As is the case with other therapeutic techniques (e.g., in psychoanalysis), the importance of narrative creation can take precedence over documentary accuracy. If the "sense" of documents in their social use is pragmatically workable, then the criteria for referential accuracy in the documents can be "good enough," rather than exact. (Indeed, documentary accuracy usually follows this rule; for example, a general encyclopedia's measurement for the Golden Gate Bridge may be correct enough for a schoolchild's paper, but not for an engineer planning on repairing the span. Correctness is always in regard to some measure and some judgment of measure, and also there are various types of conditions for truth that are valid for describing different types of facts, such as correctness, completeness, timeliness, etc.) This doesn't mean that information professionals shouldn't strive for the best documentary accuracy, at least, in the above case in order not to "revictimize" the families of the disappeared (Medina & Wiener, 2016). Rather, I am suggesting that the notion of truth in the right to truth is not solely reducible to documentary accuracy and that there are therapeutic criteria in narrative involved, as well.

As Naqvi (2006) argues, the notion of truth in reconciliation commissions attempts to address the past in terms of a desired future, as well, where the truth serves not just a personal psychological, but also national reconciliatory or judicial purposes. The information demanded within such a right, therefore, can depend on a larger context of need than a simple request for a certain document or even documents on a specific topic. The right to truth is oriented toward a broader context of justice, not only in the present in regard to an event in the past, but also toward preventing abuses in the future by having access to the facts of what happened in the past. For this reason, this "right" is sometimes taken as more fundamental in regard to unalienable human rights than rights to information, which governments can contest for reasons of national security (Naqvi, 2006).

Governments can also view information access toward a right to truth as leading to their destabilization, and thus it has been contested by government agencies that have a stake in preserving such (e.g., the refusal of the US Central Intelligence Agency to reveal their practices of "extraordinary renditions" [Navqi, 2006, p. 266]). Conflicts between rights to truth and limits put upon rights to information, as well as political decisions such as amnesty laws and their effects upon rights to truth, keep the concept and the practice of the right to truth contentious at country-level political implementation and international law (Bishop, 2011).

As I have been suggesting, all those qualities that make the right to truth so appealing and take it beyond requests for identifiable information also make it a difficult legal concept and make it difficult to implement in an impartial manner. As Navqi (2006) intriguingly concludes, "The right to the truth stands somewhere on the threshold of a legal norm and a narrative device" (p. 273).

As we have seen in previous chapters, narratives can index and describe powerful particulars as specific causal agents across time and place, and conversely, the narratives produced by powerful particulars can model the contextual conditions of their expressions over time. These general conditions are not mere supplements to the particular, but rather they are intrinsic to the perspective brought out through the particular's powers of expression. Generally, it is the social condition of the emerging bourgeoisie of the early nineteenth century in France that Flaubert brings to light through the agency of the characters and other particulars in his novel *Madame Bovary*. Analogously, the narratives of the right to truth allow the particular to shine, but it is the victim himself or herself and the survivors that bring this light upon the matter. In this way, as will be the case with the examples throughout this chapter, rights are not merely due to the narratives and semiotics of ethics, science, and law, *defining or granting those rights*, but due to the ontological powers, and rights of powers, of agents to force themselves into ethics, science, and law. Indeed, these disciplines are founded upon *the epistemic right of agents to be empirical*. Though these are ancient disciplines, their modern foundations, discourses, and cases are historically unique in this manner.

As a narrative device, the right to truth is subject to conditions of interpretation. Whereas the right to information often involves a demand for the reproduction of documents, the right to truth results in the question of "whose truth" this is evidence of. It presupposes documents of information that can answer the need for truth, but the right to truth can also be limited by the evidence of those documents so that the demand for truth can be changed, modified, or ended, and with each the information needs may also shift or prove difficult, problematic, or impossible to fulfill.

A case of this last may be found with rights to truth in situations of prejudicial or limited documentation, as well. We can find a case like this in Brazil, in regard to the report of the public prosecutor Jader de Figueiredo Correia in 1967 regarding abuses by the Indian Protection Service (SPI),[1] an organization that was to protect the Amazonian Indians of Brazil, but instead was sometimes involved in their rape, torture, murder, and genocide on behalf of agricultural and mining interests. Shortly after its completion, the report was thought to have been burned in a fire at the Ministry of Agriculture, but then in 2013, it was rediscovered at the Indian Museum in Rio de Janeiro (Canêdo, 2013; Watts & Rocha, 2013). The difficulty of bringing this document to truth is multiple: first, its supposed disappearance; second, its continued interests in suppressing it; third, the passage of time and the death of witnesses who could support its findings; and fourth, and perhaps most of all, the political and documentary relation of native peoples to the state of Brazil. The "right to truth" is a legacy of modernity, and with that, it is also inscribed by the customs and traditions of documentary evidence of the West.

As I have been arguing in this book, documentarity is the theory and practice of Western metaphysics in inscription or "writing," and an aspect of this may be what Walter D. Mignolo calls the "coloniality of knowledge" (Mignolo, 2006). For Mignolo, modernity and the founding of modern Europe began with the Spanish and Portuguese conquests of the Americas and the plundering of its riches and the subjugation of its native peoples. This was the dark foundation for the later appearance of "Europe" and European identity and the emergence of the European Enlightenment and modern thought. It is the underbelly of modernity that carries with it the prejudices and scars of its victims in its very notion of human rights and the very concept of national and social identity (Mignolo, 2003, 2006, 2011). As Mignolo writes,

> The decolonial shift begins by unveiling the imperial presuppositions that maintain a universal ideal of humanity and of human being that serves as a model and point of arrival and by constantly underscoring the fact that oppressed and racialized subjects do not care and are not fighting for "human rights" (based on

the imperial idea of humanity) but to regain the "human dignity" (based on a decolonial idea of humanity) that has and continues to be taken away from them by the imperial rhetoric of modernity (e.g., white, Eurocentered, heterosexual, and Christian/secular). (2006, p. 313)

One of the most significant aspects of the Brazilian SPI case discussed above is the relation of the right to truth to the notion of the "human" in human rights. From the colonial period until today in Brazil, this involves the problem of how to name and treat "isolated" or "primitive" indigenous peoples within the context of modern nation-state, international laws, and human rights (e.g., to "preserve" them through "isolation" or through limited or full contact). Such peoples can remain borderline human subjects with their rights protected by nation-states in inverse proportion to their objective status as "quintessential" humans within a "state of nature." Their status in regard to truth depends not only upon written, but also other inscribed and oral evidence, which may not document their powers or oppressions.

Indigenous peoples across the world are not only brought into the mechanisms of capital and nations, but also, like nonindigenous peoples in colonially drawn national states, as well, must negotiate with such, and often find a group identity that was previously absent through their engagement with the logics of identity and difference drawn on behalf of colonial powers, their devices and centers of calculation, and the cultural forms and social and economic forces of dominant political economies (e.g., taking on broader tribal or ethnic identities than their ancestors). Still further, isolated or "primitive" tribes are brought into such colonial administrative grammars for the purpose of their paradoxical management as nationally included elements of exclusion within the state and within modernity, taken as "objects" of rights of nature.

Following Mignolo's argument, perhaps a broad incorporation of narrative techniques of the right to truth might enlarge the possibilities for documentary evidence toward understanding peoples and natural beings previously considered not to be "civilized" (by dint of writing, or at least alphabetic writing). It might do so by allowing for imaginative reframings of "information" beyond traditional documentary forms for evidence.

Rights of "truth" have to take account of the expressions of those who—by documentary techniques and the very metaphysical and ideological notion of categories of types as cases for evidence—have had their powers as powerful particulars denied. In other words, the notion of truth would have to change from that supported by documentarity—as both documentation practice and metaphysical ontology—to narrative structures that take into account the truth of beings more broadly than those in the past who have been considered fully human and so granted natural rights.

Another case from Brazil illustrates this problem in the context of the historicity of documentation and documentary techniques and inscriptions. It is the case of the Comissão Estadual da Verdade da Escravidão Negra no Brasil da Ordem dos Advogados do Brasil o Rio de Janeiro (OAB-RJ) (State Commission on the Truth of Black Slavery in Brazil, of the Brazilian Bar Association of Rio de Janeiro), headed by lawyer Marcelo Dias.[2] Here, the issue is that of trying to find resources that document the existence of people who were not considered to be, or were considered to be lesser, humans and so were excluded from creating, particularly "fixed" (e.g., written) documents as testimony to their own existence, intentions, and hopes. As in the case of slaves' narratives, poems, and other written materials in the history of the United States, the very mechanisms of writing, written language, and genre expression in the owner's language had to be learned by the slave author. The very fact of documenting one's self meant turning one's self into the inscriptions of the owner class.

The truth commission on slavery in Brazil also has to grapple with the absence or loss of records for the 350-year-old history of slavery there. While modern truth commissions commonly make use of forensic materials, these are tied to birth certificates and other such bureaucratic documents that help establish individual identity and give context to anthropological or forensic evidence, so as to help the "truth" of documented subjects appear. Such may often be lacking in the case of slaves, other than the individual identities that they had as named devices in the use and exchange economies of their individual and collective owners.

Animal Rights and Rights of Nature

It isn't with a lack of tact that we now transition from issues of documentarity, human rights, and indigenous peoples and slaves and engage the rights of nonhuman animals and rights of nature. Rather, we transition to these issues in recognition of the historical problem of natural rights being seen fundamentally as human rights, and human rights belonging to those with certain markers of human powers of expression, namely writing and, moreover, documentation and documentary evidence, understood in certain cultural forms and social norms for their deployment. Further, the argument of this book is that these documentary conceptual, institutional, and technical devices are part of a metaphysics of documentarity whose heritage has developed as "the West," not least due to the power and dominance brought through these devices, both upon the inhabitants and non-inhabitants of the West as these devices of what would become European and then modern civilization increased their scope and range.

We may recall here the management devices of current and much older "computer centers" or "data centers" (as we have examined earlier, Latour's "*centres de calcul*"), which, in the library and information science traditions, go back to antiquity and include libraries, archives, museums, and government and private record management centers. These institutions and their techniques allow for the management of humans and nonhumans across time and space. In both Latour's (1996) work and Briet's (1951, 2006) work, we have seen descriptions of the management of the "new world" by European *centres de calcul* vis-à-vis drawings, maps, ontologies, taxonomies, technologies, and institutions of representation. The ontological abstraction of entities and their institutional transformation into representational essences toward being resources for productive management is the *a priori* necessity for modern science when it is understood as engineering.

"The Animal"

The Western philosophical tradition makes a clear distinction between human animals and nonhuman animals in regard to their abilities to *represent* through language, that is, to abstract and create abstract concepts from immediate perception. Based on this difference, differences in generative powers and rights are asserted, as human beings are seen as having substantial abilities to abstract through representations their present and past experiences and reason from these to future experience. Moral rights, both obligatory and agentive, are asserted through having such an ability, as well. Human powers through cognitive skills and *techne* have also been taken as rationales for rights of control, and for assigning categories of what is considered to be the master and the servant, the human and the nonhuman, subject and object, technology and mere tools, knowledge and the objects of knowledge, and civilization and noncivilization. As we have seen throughout this book, epistemological assumptions carry with them ontological categories for both the agents and objects of their claims, but they also carry moral and other forms of practical reason for giving and exercising responsibility and power.

Jean-François Lhermitte, in his book *L'Animal vertueux dans la philosophie antique à l'époque imperial* (Lhermitte, 2015), argues that in Aristotle's works, human and animal sensibility or perception (αἴσθησις) is differentiated according to the former's relationship to λόγος, so that perception is a "full perception," that is, simultaneously being a sensual representation (φαντασία αἰσθητική) and a considered or deliberative representation (φαντασία βουλευτική) (p. 165). Translated into English, Lhermitte writes,

> In sum Aristotle makes a distinction between: a primary form of representation, one of a purely perceptual quality (φαντασία αἰσθητική), which is a lower form of imagination, proper to animals, devoid of logical or deliberative properties, and a second, a more elaborate form of representation and proper to human beings, which overflowing the φαντασία αἰσθητική, constitutes a superior form of imagination, which is endowed with logic and deliberation and thus permits one to "form a single image from many parts," that is, an image from a syllogism. The first φαντασία conserves, more or less well, in the living being what is present in reality and gives to it a certain quasi-presence. It affords that all living beings are able to have dreams, reveries, a certain form of memory, a thin form of "thought," which makes up animal behavior. But the second is a superior φαντασία, auxiliary to "full action"(πρᾶξις), and it is proper to human being, thinking intellect, and that which guides it.
>
> Thus, φαντασία clearly marks the frontier between the psychic faculties of man and animal. If there does exist an animal φαντασία it is not of a logical nature, such as the human φαντασία. For Aristotle, the nature of animals does not extend much further than sensations of pleasure and pain. Certainly, habit apparently gives a limited cognitive function to their sensations, which go beyond simple sensation. But even if some of them have access to sensation, animals do not have access to moral values. They cannot be said to be temperate or intemperate other than in a metaphorical manner, and the virtue of a horse has nothing in common with the virtue of a human being. In effect, the animal lives by sensation; only human beings can attain moral perfection (ἀρετή). (pp. 166–167)

Lhermitte goes on to argue that later Stoicism continued this line of thought in terms of οἰκείωσις ("appropriation" [see also Sorabji, 1993]), as well as in terms of the cognitive capabilities of humans and animals. The φαντασία (*fantasia*) of the animal is said to be oriented toward sensations or feelings (αἰσθητική, *aesthetike*), which allow them a certain type of knowledge of the quality of things (greenness or whiteness, for example), but animals cannot reason about these in terms of building representational concepts based on such experiences (green or white, for example) (Lhermitte, p. 167). As we can see here, ancient Greek authors and the philosophical tradition since then have distinguished human and nonhuman animal beings on the basis of two interrelated human qualities corresponding to the "higher" levels of human imagination or *fantasia*: complex language use and the ability to form concepts from such—and thus, the ability to form inferences about the likelihood of future events.[3] "Moral perfection" is a function of this inferential process. Any animal can respond, but only so many can be responsible toward the future based on inferred concepts. The failure to do so marks human moral failings.

With this observation, we return to categories and other modes of documentarity as seemingly transcendental practices of understanding. Categories, in a manner similar to writing, "fix" understand and allow inferences across different cases. The seemingly transcendental qualities of categories and writing, their seeming ability to contain different examples, gives to them not only epistemic claims of transcendental truths, but also moral claims of judgments across different examples over time. In an ironic twist, the Platonic notion of concepts as representational images returns us to the aesthetic mode of representation that Aristotle saw as fundamental not just to human beings but to other animals. Platonic metaphysics, as perhaps distinct from Aristotelian metaphysics, is a metaphysics of conceptual "images" (i.e., stable or organic wholes from parts); it is based on the imaginative power of inferred universals—a third level of fantasia, after the deliberative.

Heidegger, Animals, and World

The idea that nonhuman animals lack deliberative *fantasia* or representations has continued through more contemporary philosophy, as well. McNeill (1999), for example, argues that for Heidegger both Plato and Aristotle's understandings of cognition were structured by a notion of the governance of perception from a soul—a being that is transcendental to a body's perceptual apparatus. The eye itself doesn't see, but rather, the eye sees as part of the human organism. This seems to be a species application of Kant's notion of transcendental apperception. Within Heidegger's phenomenology, this type of view leads to Heidegger's famous and contentious understanding that animals are "poor in world" (Heidegger, 1995), which we will now discuss. The argument is that animals share our primary perception of the world, but they lack the ability to represent such via a cognitive apparatus that is transcendental to perception in the "secondary" manner that Lhermitte discusses.

In Heidegger's 1929–30 lectures, published in English as *The Fundamental Concepts of Metaphysics: World, Finitude, Solitude* (1995), he asks how we can distinguish between stones, animals, and human beings in terms of our ability to transpose ourselves into each of them. Heidegger argues, "Being transposed into others belongs to the essence of human Dasein" (Heidegger, 1995, p. 209). This imaginative transposition—this *fantasia*—Heidegger argues, is not an issue of empathy (which assumes "man is first of all an isolated being existing for himself" [p. 208]), but rather a part of our shared world of being-with (*Mitsein*) one another in the mode of human existence (*Dasein*). For Heidegger, what we share with animals is a "going along with," that is part of our being-with them and with others. This is what Heidegger refers to in *Being and Time* as "concern" or "care" (*Sorge*). This being-with, however, is with a being—a pet, for example—who does not share their human's mode of existence because it lacks the categories of this later, such as "meal," "stairs," and ultimately "world." The argument isn't that animals don't inhabit worlds, but rather that they don't have a concept of categories for "the world" and the things in it:

> Let us consider the case of domestic animals as a striking example. We do not describe them as such simply because they turn up in the house but because they belong to the house, i.e., they serve the house in a certain sense. Yet they do not belong to the house in the way in which the roof belongs to the house as protection against storms. We keep domestic pets in the house with us, they *"live" with us*. But we do not live with them if living means: *being* in an animal kind of way. Yet we *are with* them nonetheless. But this being-with is not an *existing-with*, because a dog does not exist but merely lives. Through this being with animals we enable them to move within our world. We say that the dog is lying underneath the table or is running up the stairs and so on. Yet when we consider the dog itself—does it comport itself toward the table as table, toward the stairs as stairs? All the same, it does go up the stairs with us. It feeds with us—and yet, we do not

really "feed." It eats with us—and yet, it does not really "eat." Nevertheless, it is with us! A going along with . . . a transposedness, and yet not.

. . . From the side of the animal, what is it that *grants the possibility of transposedness and necessarily refuses any going along with*? What is this *having* and yet *not having*? . . . Only where there is having do we find a not-having. And not-having *in* being able to have is precisely *deprivation,* is *poverty.* Thus, the transposibility of man into the animal, which again is not going along with, is grounded in the essence of the animal. And it is this essence which we have attempted to capture with our thesis concerning the animal's poverty in world. To summarize: the animal intrinsically displays a sphere of transposability, and does so in such a way that man (to whose Dasein a being transposed belongs) already finds himself transposed into the animal in a certain manner. The animal displays a sphere of transposability or, more precisely, the animal itself is this sphere, one which nonetheless refuses any going along with. The animal has a sphere of potential transposability and yet it does not necessarily have what we call world. In contrast with the stone, the animal in any case does possess the possibility of transposability, but it does not allow the possibility of self and another. The animal both has something and does not have something, i.e., it is deprived of something. We express this by saying that the animal is poor in world and it is fundamentally deprived of world.[4] (Heidegger, 1995, pp. 210–211)

As the above quote shows, this question of whether or not or to what degree (nonhuman) animals can go along with us and us with them, whether we share a world with a cat or dog when they share a home with us, and who it is, or whose being it is—animal or human being—who "refuses" this going along with one another—is filled with hesitations. The hesitations perhaps belong to all "goings along with," including human-to-human relationships, but it is the very ability to be *Mitsein* within a *Dasein* or a mode of existence that is at stake here. But conversely, this is also to say that a right to expression, and even a right to life—a right to existence—belongs to the modality of an existence, and so the dog's own existence depends upon its "rights" as inscribed by its being within a mode of domestication to humans. And that domestication involves for both humans and nonhumans, the very concept of "world."

"World" signifies, for Heidegger, not only the ability to go along with, but also to representationally conceptualize that going along with and the objects and relationships that belong to it. It also means to do so outside of the domains, or as Deleuze and Guattari put it, the "territories" (Deleuze & Guattari, 1987) that the animal in question has as its own within its specific existential ontology or *Dasein*. Whereas we can expect that a human being can represent their illnesses and to some extent (at least through others) their death, the Western philosophical tradition (both as philosophy and as a cultural tradition) generally precludes or limits such in the case of non-human animals. Indeed, in that tradition it is the ability to make representational concepts and infer from those that characterizes the human, and as we have discussed, the human concept of moral being and action.

Like Briet's (1951, 2006) discussion of documentation not representing a stone in a stream or the star in the sky, but the stone in a collection and the photograph of a star, the entity becomes open to having a human "world" based on its ability to conform to the "world" as itself is understood as a representational concept. The conception and experience of "the world" as a concept or representation of "our" world marks the boundary between the human and the animal, according to this tradition.

As Calarco (2015) argues, however, an animal ethics that functions through an ethics of identity is limited to only those higher-order animals with which humans can identify, which means, ultimately, that only those animals which can to some degree "go along," with humans, or (more specifically in terms of power), which humans force to "go along" with them, are granted human care.

Heidegger approaches the matter of human specificity via human difference; he proposes a documentary sense of identity for being human, which is displayed in terms of a representational ontology, despite Heidegger's critical engagement with philosophy's metaphysical tradition. Rights are "given" to animals based on their ability to "go along with" human beings in their management of the human or animal environment or by providing zones of safety for either being. And while Heidegger doesn't reduce the domestic animal to "my little doggie" status (cf. Deleuze's distaste for domestic animals in "A" of his *Abecedaire* interview [Deleuze, 2012]), still, the "going along with" concept is intimately tied to the domestic nature of a domesticated dog (or maybe cat), rather than to a flea or tick. Rights of protection and of "friendship" are given to the pet in ways that are not at all given to other animals. (Indeed, what we sometimes call "animal lovers," in this sense, often happily feed their animals other killed animals, namely those animal types that can't seem to "go along" with the representational peculiarities of the human world of *Dasein* or privileged cultural modes of *Dasein*.) A flea may have a territory, but a dog with a human being is said to share a world (as well as having its own territory), even if it doesn't have a world to its own self—that is to say, ontologically, ethically, and legally, *of its own right*. Having territories is not enough to have rights. For Heidegger, rights come with having, or at least sharing, a world, and conceptually, only humans have worlds.

Rights, as emerging from human rights, demand a sense of world in order for entities to have such. (Perhaps this is what is behind the notion of natural *bodies*, as we will examine later in this chapter, having rights— i.e., an ecology appears as having a *cohesive world* because it conforms to a Romantic organicism.)

Human rights—either innately held or given—seem to demand conceptual wholes of entities, in terms of individuals, classes, systems, or ecologies. They demand the ability to conceive of worlds as objects of our, or others,' habitation. As in Heidegger's example, rights are given for those animals that can "go along with" human worlds and the ways of human worlding. Too much animal territory and the animal can come into conflict with the human world. Too much biological reorganization in too fast of a time, as in microbiomes, and it becomes difficult to see it as human individual. Too fast of evolution, like with cancer mutation, and it becomes an object of control and in the case of cancer, war.

Heidegger's analysis of animals suggests that by virtue of their own powers nonhuman animals lack at least those rights that are common to humans. To have those rights, they must be brought within human worlds, which means for Heidegger, that the very concept of a world must be extended to them. Remembering the documentary manner of how an unknown animal becomes recognized as a new species of antelope in Briet's *Qu'est-ce que la documentation?*, that is, by virtue of their being brought within ontological categories and taxonomies, we can recognize that this struggle between granting nonhuman animals rights or not based on their own powers parallels the types of documentary processes and technologies of inscription we see proper to recognizing the being of entities. Heidegger's understanding seems to remain within a philosophical register of documentarity, in the sense that the animal is granted a world only within the Dasein of human domesticity. As we will see, in modern rights theory, however, we see nonhuman animals increasingly being given rights, at least ostensibly, based on their own powers (though, also as we shall see, the concepts of domesticity, organicism, and worldliness haunt these quasi-empirical recognitions of nonhuman animals, as well).

Semiotic Being

Whatever the representational qualities of Heidegger's analysis, his work also suggests the specificity of nonhuman modes of being and suggests that such specificity or uniqueness can be extended beyond species modality and toward individual selfhood for animal entities. We may then ask how can this be done in terms of rights, if we don't want to revert to an ethics of human identification (whether one based on psychological identification—such as empathy or sympathy—or on ontological qualities)? Is there a way of re-territorializing the relation between human and nonhuman animals so as to rethink the notion of not only "animal," but also the rights of nature, and so with this, natural rights as extended to nonhuman entities?

One way of trying to think of the specificity of other animals, not in our world, but within overlapping and mutually conjoined territories, would be by prioritizing the indexical relations of entities *prior to* (or independent of) their class identification. As I will examine below, doing this, for example, within a semiotic register of indexicality, might also allow us to view language, qua tools of communication, as a property of all beings in relation to one another and their environments according to a principle of general indexicality rather than class relations. We could call this a type of "environmental semiotics." It would be a communicational, rather than a documentary, indexicality for being-with.

What I'm calling an "environmental semiotics" begins with the integrity of the entity from the aspect of its specificity as a powerful, expressive particular within its own communicative ecologies, rather than its belonging to human worlds via documentary categories or its ability to "go along" with certain representational qualities of human existence as, say, a domesticated animal.

Eduardo Kohn offers an argument for such a semiotics in his work on the Runa people and their natural environment in Ecuador's Upper Amazon (Kohn, 2013). Almost stereotypically, as is the case with many such anthropological stories, it starts from an "indigenous" perspective. In Kohn's telling of the Runa's relationship to their environment, the human is included within the expressive powers of their natural environment, rather than the reverse. The very notion of the "natural" is not distinct from the "human," but rather the set of natural entities include human beings, and in Kohn's book they are inscribed together within a Peircean type of semiotics of inter-entity and interspecies communication. Kohn doesn't do without types or kinds, but rather, he argues that categories emerge as expressions of selves in relation to one another in an environmental situation. Writing about the Runa people and their relation to an "ecology of selves," Kohn writes,

> A focus on this living semiotic dynamic in which indistinction (not to be confused with intrinsic similarity) operates also helps us see how "kinds" emerge in the world beyond the human. Kinds are not just human mental categories, be these innate or conventional; they result from how beings relate to each other in an ecology of selves in ways that involve a sort of confusion. (2013, p. 16)

But what is a self for Kohn? For Kohn, it is an agent that is a product of a "semiosis"—a communicational ecology of signs within which the being finds itself:

> Wherever there are "living thoughts" there is also a "self." "Self," at its most basic level, is a product of semiosis. It is the locus—however rudimentary and ephemeral—of a living dynamic by which signs come to represent the world around them to a "someone" who emerges as such as a result of this process. The world is thus "animate." "We" are not the only kind of *we*. (2013, p. 16)

Thus, the self is an indexical point of communicative or affective relationships, past, present, and with trajectories into the future, that has expressive agency. Deploying Peirce's vocabulary of "iconic" and "indexical," Kohn writes,

> Indexicality, then, involves something more than iconicity. And yet it emerges as a result of a complex hierarchical set of associations among icons. The logical relationship between icons and indices is unidirectional. Indices are the products of a special layered relationship among icons but not the other way around. Indexical reference, such as that involved in the monkey's take on the crashing tree, is a higher-order product of a special relationship among three icons: crashes bring to mind other crashes; dangers associated with such crashes bring to mind other such associations; and these, in turn, are associated with the current crash. Because of this special configuration of icons the current crash now points to something not immediately present: a danger. In this way an index emerged from iconic associations. This special relationship among icons results in a form of reference with unique properties that derive from but are not shared with the iconic associated logics with which they are continuous. Indices provide information: they tell us something new about something not immediately present.
> ... What is the relationship of indices to symbols? Imagine learning Quichua. A word such as *chorongo* is relatively easy to learn. One can learn that it refers to what in English is called a wooly monkey quite quickly. As such, it isn't really

functioning symbolically. The pointing relationship between this "word" and the monkey is primarily indexical. The commands that dogs learn are much like this. A dog can come to associate a "word" like *sit* with a behavior. As such, "sit" functions indexically. The dog can understand "sit" without understanding it symbolically. (2013, pp. 52–53)

Kohn sees meaning as originating from and extending further out into the world than human language's "iconic" or representational symbolism, and he argues that forests "think" as ecosystems through indexical relationships. As we see from the last few sentences in the quote above, in contrast to the Western philosophical tradition where thinking and language often connotes representation, for Kohn thinking and language are indexical forms of creating reference—in this sense, not so different from Latour's map reading, but for Kohn inclusive of all entities in the world, including humans. Evidence is gotten through associations between words and things in experience. Reference is the product of these indexical relationships of sense, not the prerequisite for such. The "self" in such a schema is an index of experiences in the present, simultaneously pointing to the past and future. Living beings, for Kohn, are intrinsically joined together, because they are formed and continue to grow from out of connected systems of indexical signs that are "pragmatically" understood. In a manner, the ecological lifeworld that Kohn sketches is made of networks or graphs with indexical relationships between the organic points making up the edges.

Since, for Kohn, the entire lifeworld is alive with meaningful events that are registered in the subsequent relational expressions of each being, the notion of the self is extendable even to plants. And for this reason, as well, Kohn claims that "forests think." For Kohn, forests are sort of environmental brains, alive through connected experiential synapses. "Aboutness" is a product of ecologies; it is product of semiotic affects, not *a priori* categories:

We humans, then, are not the only ones who interpret the world. "Aboutness"—representation, intention, and purpose in their most basic forms—is an intrinsic structuring feature of living dynamics in the biological world. Life is inherently semiotic. (2013, pp. 73–74)

By extending Peirce's semiotics to beings as a whole, and giving to entities the notion of a "self" based on its undergoing and adapting to affects (akin to Whitehead's process philosophy), Kohn give us a sort of semiotics gone wild, where communication and knowledge formation fundamentally occurs between human and nonhuman entities without distinction.

Rights of Expression 129

For Briet, the animal entity is an indexical sign because it belongs to documentation ontology; in contrast, for Kohn, it is an indexical sign because it is a self within an ecological semiotics. Instead of nonhuman animals being mute, it is the human beings of the Western cultural tradition who may be seen as being both deaf and mute to the language of their environments, captured as they are within the domesticating worlds of their representational signs. Humans of the metaphysical tradition narcissistically turn away from shared semiotic networks with other animals and plants; they fetishize their selves through metaphysics of transcendental identities, to which they see themselves as the highest form. They are captive to the power of their imaginations, blinded within their *fantasia* of representational knowledge and transcendence.

Kohn's (2013) critique of an anthropomorphic view of language and cognition makes an impassioned argument for an epistemic and ontological model of "nature" and indigenous life based on a general semiotics of affect. Kohn's argument suggests that by privileging a transcendental understanding of "aboutness" or representation, the sensual foundations of such in environmentally determined indexicality are lost sight of. Sense is lost, and indexical reference is then transformed into transcendental representations.

As Briet briefly mentions in the first part of *What Is Documentation?* (2006), the techniques and technologies of documentation bury or veil the original documental fertility of the newly discovered animal. Whereas such documentation techniques or technologies veil the animal in the name of its type essence, Kohn's work (2013) attempts to reopen the ontology of beings by an environmental semiotics that stresses the singularity of each entity.

However, one might also object that the very generality of such an environmental semiotics can also obscure the problem of the specificity of beings as entities, and even as class, particulars. Kohn's semiotics ontologically grants expressive agency to nonhuman animals, but it does so at a very general level of sensory response, thus perhaps reducing entity powers to that of system responses. Semiotic indexicality can model expression as response, but can it also model expression as integral to the experience-built ontologies of individuals, either as a species or as singular beings?

Earlier in this book, we saw an expressionist theory of singular and individual agency in terms of dispositions and affordances. Following this, I'd like to turn our attention to recent attempts to not only give to nonhuman animals, but also to "natural" or environmental bodies, rights of expression. Such rights are both expressions of dispositions and, when viewed against strong documentarity and the documentation tradition, also rights to *not be* an entity as conceived by human knowledge. In this latter, they can be seen as rights to *not* "go along with" human *Dasein*—not to be domesticated or to fit within modes of human representational governance. In a sense, they are rights to be or remain "nature," or "wild." Whether such rights are even possible remains not only a pragmatic issue of legal processes, but also a conceptual issue of just how far into natural beings and objects "human rights" can be extended. Can such rights exist in the absence of their conception as rights by human owners of such? Proponents of rights of nature claim that they can be, that natural bodies have rights of expression against human claims, even if it is humans that must press for those rights through legal institutions.

From a contrary angle, though, we could also argue that natural entities have, at least *expressive*, rights *not* to be necessarily included in the worlds of "human rights." Indeed, the natural sciences imply such, in so far as they attempt to investigate that which is not available to ordinary human experience, using tools of investigation and interpretation of the others' nonhuman powers. Granted, these latter are not contingent on rights of continued existence, but rather, rights of claims of expressively independent existence. As Deleuze (2012) argued, territories belong to the flea or tick, or even to the dog, but worlds not necessarily so. The concept of a world doesn't ontologically belong to a dog, though its domestication lends it one and may make its existence more long-lasting; territoriality, however, belongs to it ontologically, whether it is domesticated or not. Ontological rights for entities may not even need something like Kohn's (2013) semiotic world of nature, though legal rights may.

Animal Rights and Rights of Nature

If the development of modern "natural" (i.e., human) rights of expression and information access have been disruptive to taxonomies of being and power as embodied in obligation or duty rights, so more recent animal rights discourses, and more recently rights of nature and earth jurisprudence discourses, show an even further "rights drift" toward extending the notion of what we can call expressive or agency rights more toward powerful particulars outside of human civil society. Each of these rights discourses has been extended from the original Enlightenment break from obligation rights, forming around a fundamental ontological understanding of individuals as powerful particulars.

This rights drift toward viewing the particular in terms of its singularity also brings with it disruptions as to what can be seen as documenting evidence of such rights, from that of transcendental properties to notions of group essences held within individuals to that of powerful particulars and powerful "particular" ecosystems. In brief, we see a gradual drift in such rights discourses from recognized essence by others to self-expression of "organic" bodies as the source for rights, and so we also find a tendency toward taking the powers of the particulars as grounds for their own self-evidence.

The drift from animal rights to rights of nature would, as it were, seem "natural." Both "animals" and "nature" are traditionally understood to deal with nonhuman beings, and so the extension of rights of humans to animals and nature together makes sense. However, the equating of animal rights with rights of nature is complicated.

First, as we have discussed above, we should recognize that animal rights often involve the extension of human rights to domesticated animals, whether those animals are domestic animals ("pets") or are captured wild animals. Within the "house of the human," as it were, we extend rights to those animals that, by choice or by force, "go along with" us. The minimal type of rights given to such animals in the past two hundred or so years, at least in Western culture, is that of freedom from pain and suffering, or to put this another way, protection from "unnecessary" pain and suffering. Of course, what constitutes "unnecessary" varies widely on who is interpreting this and how. Is a bull's quick slaughter painful, and it is unnecessary pain and suffering when compared to other ways of dying? Is euthanizing a cat with suspected cancer, saving it from the pain and suffering of a surgically produced diagnosis and possibly noncurative surgery that it cannot understand, creating or relieving unnecessary pain and suffering? Not only are culturally specific human values and particular human conveniences for measuring the animal's pain and suffering called upon to both enact and justify actions in the above scenarios, but the very measure of pain and suffering is difficult to judge on a psychological level.

Second, animal rights often conflict with one another. Domestic cats are beloved in much of the world, but cats are biological predators; they not only instinctively hunt, but they also physiologically need to eat other fleshed animals, such as their most prized prey, birds. Hence, it is not surprising that bird lovers are not always cat lovers (and within this the irony also exists that some birds are predatory and many enact infanticide). If pet food stores are for "animal lovers," then it better not be as lovers of those animals whose flesh is in the cans and bags throughout the store. Further, many "nature lovers," not least those who love wild birds, find little to love in cats, which have a negative effect (far below that of the top predator, human beings, however) upon bird populations. Thus, there are cases where animal rights in the name of "unnecessary pain and suffering" conflict when applied to different types of animals understood within human customs.

Pelizzon and Gagliano (2015), however, like Kohn (2013), suggest that plants, too, are sentient beings and so, they argue, also deserve rights. And once plants have rights, along with all animals, then it is not a far stretch to say that a forest or other natural ecosystem has rights. Such rights can be based on empathy, according to the "unnecessary pain and suffering" argument or, with slightly less anthropomorphism, as we have seen, can be based upon a semiotic or cybernetic notion of life or a life system.

Third, when we extend rights to natural systems based on animal rights, we are extending the notion of an organic body to such systems and also assuming that such systems have the ability to suffer, have pain, and have autonomy, akin to animals. The moral and legal rights given to natural systems, when viewed from animal rights, may be based on emphatic and harm-based precepts in human morality and law, but unlike animal rights, it may be difficult to see pain and autonomy for such systems as equivalent to those of individual animals.

Further, social institutions seem to import notions of organic wholes for purposes of representation in order to perform natural rights discourses in political and legal contexts. "Animals," "trees," and "rivers," taken as class and ecologically connected wholes, also need human advocates in political and legal contestations. Like conservation efforts based on the notion of nature as being a "service provider" to humans (Flint et al., 2013), such cases initially took the form of claiming grievances on behalf of human plaintiffs when ecological harm was done. (E.g., waste-dumping in a river poisoning a farmer's privately owned lake downstream.) More recently, however, in the United States, legal cases have increasingly been contested on behalf of natural entities or "bodies" (e.g., a "body" of water) *themselves* as plaintiffs (Stone, 2010). The law seems to demand the extension of the notion of subjectivity to nonhumans in order to grant them legal status, and such subjectivity includes the notion of an organic whole or "body" or "system." Such a precedent is well established in the United States in the case of corporate bodies; its extension to natural bodies is more recent. These legal extensions of self or persons to natural entities or bodies constitute "rights of nature."

Despite their being oriented toward the animal or natural "other," empathy models for animal rights and rights of nature are essentially human rights arguments. Semiotic and cybernetic narratives, such as what we saw with Kohn's (2013) work, displace these notions of subjectivity, but they may do so at the cost of enrolling all beings in "natural systems" of a very generalized sentience. Discourses and laws based upon natural rights, such as Ecuador's famous chapter 7 of title 2 of their Constitution of 2008, which gives rights to "Pachamama," and Bolivia's *Ley de Derechos de la Madre Tierra* of 2010, both start from the notion that the earth is a "mother" (which, needless to add, is a very anthropocentric concept). Further, such approaches sometimes attribute such claims to indigenous peoples, with indigenous peoples being taken as having a "naturally" closer relationship with this "mother," and so being part of the natural system, distinct from those people identified more closely with colonization.[5]

Despite the limitations discussed above, seen within terms of (eco)system theory rights of nature may be viewed as extensions of animal rights to groups of animals and plants, and even to the earth, all recognized as organic unities of dispositional powers. With rights of nature, we come to a form of rights where the expression of dispositional powers of organisms or superorganisms is the primary site of right-bearing agencies. Natural "systems" are viewed as powerful particulars, akin to how individual animals are seen as wholes composed of cells and other organic parts. Rights of expression are based on expressive powers emanating from such systems as unique, particular ecologies. To destroy the "system" is to endanger not just the animals within it, but also the whole as a complex system whose expressions are greater or different than its parts.

On the other hand, such a "systems" approach to ecologies, starting from stable micro-particulars and ending at the point of unified, macro, bodies may negate the role that evolutionary selection and "competition" has in establishing any organic body. Recent work in esophageal cancer cell biology, for example, suggests that what at a certain level may be seen as an organic body of interfunctioning normal cells with a few problematic mutations, is in reality, an ecology of not only "normal" cells, but also a very large ecology of competing mutant cells (Martincorena et al., 2018). The body that results (e.g., an esophagus) maintains its integrity because some mutant particulars lack the ability to mutate the organ(ism) in some other direction of development.

Thus, while the notion of an ecological body makes more complex the notion of a "particular," in itself, it often rests on organicist assumptions grounded in a certain level of observing entities and their interactions. Even when thinking of particulars as dynamic, we often remain trapped within organicist notions of wholes and parts, and thus representation.

In summation, in this chapter we have explored the expressive powers of particulars that lead to their being given ethical and legal rights in human worlds. Such legal rights, like scientific "rights" of powerful particulars, are due to their perceived ability to assert themselves, to be "autodocumentary," and to push back against particularly human modes of cultural interpretation. Documentary categories are said to follow, not precede, these self-expressions.

As we mentioned, however, in reality documentary categories often begin scientific investigations through taxonomies and ontologies. In scientific research, beings are ontologically parametrized, at least by nomenclature and often by taxonomic structures, and they are then studied within those parameters according to methods and techniques deemed appropriate to their subject matter. And nonhuman animal and ecological "bodies" are given respect and standing in the human households of domestic and legal representation.

As will be discussed in the next chapter, through social network algorithms and newer artificial intelligence technologies and techniques, empirical senses take on a new importance in the construction of documents, data, and evidence of all sorts, and the reuse of them in real time events. Documentary fragments in the sped-up time of Internet social communication carry with them documentary claims to truth but within a communicational, even a conversational, framework. Powerful particulars, both of social and natural entities, thus become directed by representational becomings that are governed by not only *a priori* categories of being, but also by predictive data vectors of represented habits, knowledge, and social relations. In the next chapter, I will collectively discuss such technologies as "post-documentation" technologies, which I see as technologies that attempt to not only measure, but also sometimes learn from, powerful particulars, while also shaping such particulars and their actions as documents for further representation and social and technical reuse.

7 Post-Documentation Technologies

In the previous chapter, I investigated the issue of the powers of individual entities—powerful particulars or "singularities"—to, in part, self-document themselves in their expressions, and so, within modern rights theories, to have claims of agency rights. As I have discussed, in the Western tradition we see a general historical trajectory from a "strong documentarity" to self-evident powerful particulars and then to the representation of entities by their empirical traces. In the empirical sciences, technologies, techniques, and inscriptions must construct affordances that allow for the expression of a studied entity's dispositions. Drawing the line between science as affordances that allow dispositions to be expressed and engineering, which allows only certain dispositions to be expressed toward systematic use, is not always easy to do in the technosciences. This is particularly true in the social sciences, and information science as part of such, for the dispositions of a self are socially and culturally constructed and anticipated.

This chapter looks at the problem of evidence from what I call a "post-documentation" perspective. This perspective stresses the principles enumerated below as representative of the current state of documentarity as a shift to evidence attained more on the horizon of empiricism and sense, rather than categories and reference. Post-documentation technologies can result in *computer-mediated judgments*. Algorithms are not just the documentary "techniques" (Briet, 1951) of yesterday,[1] but they are real-time technical mediations of information and communication, and with these, judgment. (Where previously in traditional documentation practice, documentary techniques and technologies were more explicit and external for judgment, now they are more implicit and folded within judgment.)

Increasingly, Rimbaud's modernist edict of "Je est un autre" becomes manifest as rapidly composed and changing judgments in real time, not only by human judgment, but also by computer-aided judgment and by computer-produced judgment. The future will increasingly be composed by computational judgments.

Post-documentation could be said to have the following principles:

1. Striving to represent particulars *qua* their powers, affects, and trajectories ("sense").
2. Particularity should be represented through particular modes of expressions proper to the individual and group types of the particular entity.
3. Modes of expression are emergent out of historical situations and situational contexts for individual particulars, which help afford their expressions and the meaning of those expressions for others. Human expressions are constantly mediated and afforded by sociocultural and technological inscriptions, though they can have some choices among which to be inscribed by, indexed through, and which to deploy in expressions. (Principle of historicity and freedom.)
4. Digital "information infrastructure" refers to sociocultural-computational techniques and technologies (*techne*) that result in meaningful or "informational" expressions (documents) that can be joined together in communicative exchanges and streams of information.
5. Reference for meaning and truth-value should be borne by sense, rather than solely by *a priori* class categories or formal parameters throughout the processing of information.
6. *A priori* categories, such as those that result from classification structures, can be heuristics for investigating entities, but they are only that.
7. Information is the product of communicative or affective relations. "Being" is a relation and a product of evolution. Being becomes evident and is recognized as evidence, but it can have evolved otherwise and could be recognized as otherwise from different perspectives and scales of evolution and analysis.
8. Reference is a product of the theories, methods, and technologies used to understand entities in relation to their powers. The subjects and objects of reference should be understood by their senses of expression, as seen within a "critical" epistemic perspective.

9. Science and scholarship are primarily sets of activities and social functions, not the product of such activities in statements, documents, or their institutions (e.g., books, libraries). These latter are expressions of science and scholarship in certain moments of their production and institutions.
10. "Document" still remains an important concept when analyzing post-documentation technologies, but in regard to these technologies documents can be seen as products of them, as much as the input for such.

The central point that I will develop in this chapter is that of a contrast between documentary representation as a product of *a priori* categories of identity and difference (i.e., the products of traditional documentation technologies) and documentary representation as more a product of *a posteriori* or empirical sense (i.e., the products of "post-documentation" technologies); that is, post-documentation technologies are less technologies of direct reference and more technologies of sense (from which reference is then taken).

Though we have seen in this book the tension between concepts of documentarity as being reference and sense driven in the genres of "philosophy" and modern "literature," respectively, I will suggest, in this chapter, that these genres have been bridged to some degree by post-documentation technologies today. Any understanding of information mediation must begin with the social-psychological pole that we can call "ideology," on the one hand, and the technical pole that mediates this into the delivery of a "need," on the other (Day, 2014). Online, we essentially live in a world of cybernetic systems, where technologies follow our expressions and position us in documentary space through data minute by minute, in small time scales (e.g., GPS location) or larger time scales (for most of us, in non-extraordinary times of our lives, Google search rankings). Like the servo-mechanisms in gunnery apparatuses during the Second World War, which exemplified the theory of cybernetics, information systems position us in order to maximize our information search "hits." Information mediation for the human social (and even nonhuman, but human-managed) world spins around this dialectical axis of data being positioned for us through us being positioned by the data. The purpose of information technologies is to produce reference out of sense traces of data, which sometimes means creating single referential documents, or more often now means creating sensual vectors for agent identities through collecting recursive inclusions from the past, indexing the present in socio-documentary matrixes, and creating predictive networks to be shaped and verified by future data. Realist genres, whether textual or visual (like game playing), further lead to both high precision and recall by limiting the domains of informational sense and reference. We should realize that when we talk of the empiricism of post-documentation technologies, we are also involving paradigms for collecting and processing data and for making judgments about such. Like empiricism in the sciences, there are, of course, preliminary ontologies and other paradigms involved and realist expectations of what are pragmatic and valuable uses for information.

Post-Documentation's Philosophy of Language

An historical account of this renewed empiricism in information science could start with the recent shift in the philosophy of language behind contemporary information retrieval. David C. Blair's works during the 1980s up through his large volume on the importance of Ludwig Wittgenstein's philosophy of language for information retrieval and information science (Blair, 2006) theoretically foreshadowed and accompanied a seismic documentary shift (at least on the Internet), from category-based indexing and searching to link-analysis, social-network, and machine learning systems. The shift from traditional documentary indexing by means of subject headings and other category or class descriptors to graph analysis was signaled in the search engine world by the triumph of Google's PageRank algorithm in Google Search over Yahoo!'s earlier directory structure.

Underlying both the theoretical and practical shift in information retrieval is the epistemological perspective that meaning and value in language are products of its use, not products of categories of the mind. Mental categories are themselves viewed as products of language and its use in the world, in relation to language and nonlinguistic materials and events. Documentary categories and other modes of traditional reference are seen as products of cultural forms in social use. In contrast, older information retrieval, emerging out of traditional documentation technologies and techniques, forced social use to follow very professionally constructed and controlled cultural forms, such as controlled vocabulary or subject headings.

Traditional documentation was based on category application to objects, and in bibliography to the content of works. Users had to consult authoritative subject headings or other descriptors, which named entities by means of controlled vocabularies that construct identity through structures of difference and identity (e.g., x is a "dog" because x is not a "cat," etc., in Library of Congress Subject Headings). In contrast, post-documentation technologies, such as social-network analyses and machine learning algorithms, give value and meaning based on statistical calculations of the use and the relations of data, including, of course, language use.

Returning to our historical account in order to provide more detail, let us recall that both Yahoo!'s directory structure and Google PageRank attempted to address the need to increase relevancy in searching massive datasets. Except in cases of searching very unique names (e.g., the dinner party guests in the old *Monty Python's Flying Circus* skit whose names are "A Sniveling Little Rat-Faced Git," and his wife "Dreary Fat Boring Old Git"), information retrieval precision and recall generally have an inverse relationship to one another. More precision results in lower recall and higher recall results in less precision.

Freely searching a database of indexed websites the size of the Internet can result in massive recall for a search term, which then leads us with the problem of low precision or relevancy. Without further algorithmic adjustments, more recall of a keyword term on the Internet generally brings about less precision. For example, in the historical beginnings of the graphic user interface internet during the mid-1990s, a search on "World War II" in AltaVista—initially, a more or less "pure" keyword search engine—brought back for the user all sorts of results mixed together without meaningful ranking for a user's query: World War II documents, souvenirs, personal remembrances, and so on.[2]

What to do to improve this? In the case of the search engine developers at the time, the answer seemed to be to look at what librarians and other documentalists did in the past when they had to organize information.

So, Yahoo! and some other companies at the time employed "ontologists" through either full human means or computer-assisted means to create categories or "directory structures" for searching (e.g., "Art>Painting>Europe>19th century>Impressionism"). However, one of the important limits with such a traditional documentary approach is that of determining the meaning of documents through names that represent the essential "aboutness" of a document. Such naming occurs through professionally managed controlled vocabulary, and so it requires consistency in naming by catalogers or ontologists, and on the other side of indexing, namely search, it requires that users know the preferred name for an indexed entity.

Google's PageRank algorithm took a different approach, in part utilizing another library tool, that of citation indexing, whose origins goes back to legal indexes in the nineteenth century and the important work in the second half of the twentieth century by Eugene Garfield in bibliometrics and his creation of the Science Citation Index (Rieder, 2012). Distinguishing it from systems such as Yahoo!'s directory structure, Google Search crawls and indexes the web and then increases relevance in search through a link-analysis system called PageRank. (There are other means for increasing relevance in Google Search, but link analysis was the distinguishing means from earlier systems like Yahoo!'s directory structure.) Link analysis is an adaptation of citation-analysis systems (which are still used in scholarly communication) for the indexing and ranking of documents on the Internet (and so, a tool for searching), which attempts to increase precision or relevance by social means.[3] Relevance is increased through algorithmic calculations of what others believe are the most important documents for subjects. Seen from the philosophy of language, the success of PageRank (and hence Google Search as well) vindicated a Wittgensteinian philosophy of language, at least when applied to a socially broad databases such as the Internet. It also established the principle that meaning, value, and categories, too, are established through *relationships* between documents and documents, people and documents, and people and people. Graph data structures and algorithms are thus an important tool in mapping meaning by the social use of language.

However, this same approach also has a downside, particularly when used in extremely large public document systems such as the Internet. The downside is that social sense doesn't necessarily lead to knowledge or better knowledge. More information is not necessarily more or better knowledge. "Knowledge," at least in the institutional sense of this term during modernity, has referred to information that has passed through institutional authority for verification and assurance as to its likely truth-value. Such means require methods of evaluation, peer review, consensus among experts, and at least potentially, challenges to the factuality or truth of claims. (Even in nonscientific institutions, such as journalism, fact-checking should take place.) Google Search indexes, retrieves, and ranks documents that are "information" in many senses of the term—opinions and rumors, as well as institutionally vetted knowledge documents. Seen from the perspective of modernist knowledge institutions (scientific, bibliographic, and so forth) Google Search is, generally, an information indexing and retrieval system, not a knowledge indexing and retrieval system. (In contrast, for example, Science Citation Index, as a scholarly communication index that indexes peer-reviewed articles in approved journals, is a "knowledge" indexing and retrieval system in the institutional sense of the term.)

Sense

Earlier in this book we discussed Paul Otlet's theoretical understanding of documentation, based on his belief that books and other documentary materials contain factual representations of the world and that classifications are knowledge organization systems. If traditional documentation is an epistemology and a set of technical tools based on creating and applying exclusive classes or categories of aboutness to entities, then what about the phenomenological senses of those entities? What happens to the documentary paradigm if the "aboutness" of an entity is determined more by its own trajectories, relationships with other entities, and the material-semiotic affordances for expression and agency, rather than by categories applied to it? In other words, what happens when sense recomposes the categories?

Controlled vocabulary of documentation metadata (e.g., subject headings and thesauri terms) have restrictive levels of sense, created by means of rigid totalities of structures for imposing identities and differences among terms. Controlled vocabulary controls variance in the senses of words, so as to produce reference. An entity is either a cat or a dog in Library of Congress Subject Headings (LCSH). There are no "dogcats" as subject concepts in LSCH. Controlled vocabulary is a sort of firm linguistic structuralism.[4]

In order to remedy this restriction of sense in the construction of reference within older documentation technologies, while still providing precision or search relevance, newer information technologies try to amplify the sensibility of indexed terms through incorporating the historical, social, and geographical uses of these terms, both during indexing and in the process of searching. Examining the neighboring location of signs to one another, their semantic relationships, the social networks and grammars of their use, perhaps location coordinates or time values, and the searches of others and the user's previous searches, newer documentary technologies use social sense to find a probabilistically better balance between search recall and precision than was available either through older documentation technologies or through pure keyword indexing.

From the perspective of philosophy of language, in his book *Becoming-Social in a Networked Age*, Neal Thomas (2018) has deeply explored sense and reference creation through "new media." Thomas analyzes three forms of graph relations and their philosophies of language operating in these media today: knowledge graphs (e.g., linked data and the semantic web), social graphs (e.g., Facebook algorithms), and predictive-analytic graphs (e.g., machine learning), all constituting "post-documentary" (what I'm calling here, "post-documentation") technologies (Thomas, 2018). Graph relations create reference by means of sense through logical and algorithmic calculations of semantic, social, and recursive data inputs.

For Thomas (2018), the key to understanding post-documentation technologies is understanding how they locate and eventually package data into meaningful referential identities and categories through different techniques of rationality (e.g., analytical reasoning, social communication, and predictive computing). In the case of personal computing and mobile devices, input increasingly includes the actions and past actions of system users for calculating likely information and likely relationships between documentary items or between data. Increasingly in the development of these technologies, the post–World War II perspective of cybernetics in information system design becomes more apparent: the "user" of a system is used by the system in order to create meaning for the user and for others (Thomas, 2012; Day, 2014). The user is a data point in a communicative feedback system that rationalizes the whole, sometimes "on the fly," and sometimes further in machine processing. In short, the user is a function within a documentary system that isn't just present for the user, but rather indexes the user based on past occurences of data and, increasingly in online systems, predicted futures of data. Post-documentation technologies maintain the perspective that has been inherited from traditional documentation systems that the user's needs are constituted by the documents available, but now they make the user a function within iterative feedback or dialectical loops that are increasingly indexed to both users' and documents' past, present, and future sensibilities via relationships, weighted judgments, and predictive learning using varieties of data types (language, geographical position, friends, time, etc.).

With post-documentation technologies, the technical structures of traditional documentary technologies disappear into the background, particularly in real time use and buried in the "black box" of learning algorithms, and the social appears as data—as a given. "Social" and "technical," "users" and "documents" are dialectical poles within an algorithmic phenomenology of information and need. Social and technical parameters and functions, as well as concepts of users and documents, disappear into the "facts" of information and experiential knowledge on the Internet. What are hidden from view are not only the technical, but also the ideological and social parameters of search, including, as Thomas (2018) shows, the philosophical assumptions working in computational approaches.

Returning to a central theme and the vocabulary in my first book, *The Modern Invention of Information: Discourse, History, and Power*, we must always remember that "information" on the Internet or whatever textual form is "dragged" from experience into what used to be called "the virtual," and back again. Texts, as virtual spaces, are potential spaces for meaning. These are "informed" not simply by other texts, but by means of experience. Search engines index and rank the real into the virtual, and these virtual spaces are actualized and made real by means of experience.

"Structured data," in classification structures, and to some extent with linked data, still maintain an epistemic distance between the semantic or "virtual" space of the text and that of experience. It is with link-analysis and other social systems that we have begun to see the greater folding of lived experience into information systems and the mutual overlapping of formerly more distinct information, communication, and media spheres.

Like with older documentation tools, such as classification structures, the danger of "the information age" is that of its seeming disappearance at the moment of its triumph; that is, the triumph of the mediation of reality by "the virtual," the internalization of what was a distant and explicit documentary tool into a necessary affordance for not only knowing, but also perceiving and actively judging, in everyday life. This is when all of reality is *media*ted by "virtual" (or to use this term in another of its senses, by potential) relations. When the text becomes real. That we no longer speak of digital information as "virtual" (and with this, as potential), and we no longer speak of "the information age," but now live it as the real and logically possible, instead of as simply potential, I think is indicative of the evolution of the information age into being time itself for us.

We earlier looked at the characters in *Madame Bovary* as examples of such an event in a much earlier age of documentation. And we saw Derrida's reference to the problem of the fable, or the "as-if" structure of texts, when they are made real as informational media. My earlier books, and now this one, have been tracing this trajectory of the realization of the modern sense of information in our present time as ordinary lived experience, and its long evolution from the beginnings of Western metaphysics until now.

Sense and Reference in the Media Sphere

The notion of "information" as belonging to the domains of documentation and fact, and "literature" as belonging to the domains of fiction and the imagination, are in part being blurred by information being detached from institutions of knowledge production and their techniques and methods of verifying information. The line has never been solid, of course, since "literature," as well as art, as we have seen, is made up of all sorts of performative activities using techniques and methods that are then, at least in personal experience, "verified." Also, human beings constantly rely upon noninstitutional information, treated as knowledge, to get through their daily lives. Personal knowledge and experience are, of course, very important in our lives.

However, despite these issues, there remains the problem of the blurring of models of literary or aesthetic reality meant as analogical road maps for personal experience with institutionally mediated knowledge. Modern institutional knowledge arose in order to address problems with speculative, theological, and purely personal, experiential, knowledge. In our present day, this blurring (to the detriment of institutional knowledge) has to do with political conditions of resentment, distrust, and skepticism toward "official" institutions, and also with the availability of, and constant daily mediation of our lives by, noninstitutional information of many types.

Institutionally mediated and produced knowledge is always going to cost—in the time needed to produce, understand, finance, and preserve it. Libraries, laboratories, and universities are all parts of knowledge construction and circulation systems, though they do, of course, handle information of other types. They can be costly to run. Information that everyone already more or less knows, of course, is likely to be cheaper in all these aspects.

Because communication now occurs in published forms in social media as short documentary assertions (e.g., Twitter tweets and alike), we sometimes assume that the rhetorical and institutional knowledge construction procedures and modes of circulation that have existed for documents exist for these documentary fragments, as well. We may treat these fragments or "memes" as the conclusions to documentary enthymemes. We can be fooled by the evidential trust that we have learned to put into published texts. And when these fragments are then algorithmically used to generate more fragments or generate networks of fragments or fuller documents of fragments, what we then may encounter are networks and constellations of assumptions, hysteria, and innuendo, where the agreement of other "speakers" lay credence to the truth of the assertions being made.

As I write this, the *fait accompli* of the merging of fiction and documentation is being widely discussed: Phenomena such as "fake news" and political memes have had tremendous political effects. And while false information in the modern media is nothing new, what is new today is the speed and vast participation in live-time feedback systems that are merging communication and documentary ecologies into lived "information."

Today there is a mixing of documentation, communication, and "media" ecologies, all with claims of being "information." What is lacking is asking how they are each "informative" and if their informative elements are knowledge? And if they are knowledge, what are the manners and criteria by which claims are validated?

A "post-truth" era shouldn't be the same as an era of falsity, but rather, it should be an era of the true. It should be an era of understanding different types of truth claims and their rhetorical and social forms, including identifying flatly false assertions.

As I discussed in *The Modern Invention of Information*, Walter Benjamin's works of the 1920s and 1930s comment upon a similar dynamic of new media forms (cinema and radio) of "information" in Weimar and Nazi Germany as what has occurred with the Internet during the past thirty years. Then (both in Germany and elsewhere, such as in the US) and now, trust in mainstream institutions and liberal state political parties eroded due to economic distress and political corruption, and along with this the fragile foundations of the modern democratic state, lying in a trust in modern institutions and bourgeois cultural habits, dissolved.[5] In the light of the appearance of "new media," together with distrust of social, cultural, and political institutions, modernist knowledge institutions were cast aside and "the facts" of information were stated through new information technologies. New media doesn't automatically lead to the erosion of modernist knowledge institutions and the resurgence of prejudice, but it opens up the space for new kinds or sources of information at lower economic and transactional costs. As Benjamin observed in the case of film (Benjamin, 1968b), at first this can be quite liberating for people who are trying to gain information, as the technological and institutional hindrances seem to be, at least at the point of searching, removed. And new media may focus (literally, in the case of film) upon events that were little depicted earlier, making available and magnifying that which escaped older media and knowledge institutions. But then this information space often becomes remediated through the return of old media and old social prejudices within that space.

The problem of the circulation of the documentary fragment in media and its reconstitution as a document by ideology and social prejudice well illustrates the increasingly "sensual" nature of documentary "aboutness" or reference, the shift to social senses, such as taste, for information, and yet the carryover of the aura of documentary knowledge. We expect our published information to be, if not true, then at least potentially truthful, or at least potentially arguable, and there to be evidence. Social sense, however, is not limited to this. Much of our use of the Internet is governed by taste—by likes and dislikes.

Computer-Mediated Judgments and Machine Learning

I am not a computer scientist, so my comments here will be very brief, but this book would be incomplete if I didn't at least touch on machine learning and its construction of judgments through neural networks and deep learning techniques and technologies. It seems to me that such technologies are at the forefront of creating categories and representations as computer-mediated judgments, and so also, they suggest and generate prescribed actions that might arise from these, for both humans and machines.

Computer-mediated judgments take the form of, first, support systems for human judgments and, second, direct computer judgments. Examples of support systems are AI systems that support human activities such as

Post-Documentation Technologies 149

identifying suspected persons in crimes. Direct computer judgments are those that result in direct action following computation or chains of computational systems, such as future automated weapons systems that could identify suspects and kill them or completely automated cars. Computer-mediated judgments take place throughout the Kantian categories of judgments: those weighing data and deriving conclusions (theoretical or knowledge judgments), decision support systems that enable human and machine action (practical or moral judgments), and even the automated construction of novels and other works of art (as products of judgments of taste). Increasingly, processing results will be integrated into other processing systems, bridging theoretical, practical, and aesthetic judgments.

Directly or indirectly, already and increasingly in the future, beings will increasingly be shaped by such computational judgments, as computers continue to shape us as both active subjects and as the objects of knowledge, action, and taste. Our children no longer see themselves as the autonomous subjects that we saw ourselves as even thirty years ago, and increasingly we come to view our own being as evolutionary residues of biomes and microbiomes, reaching from the microscopic to the planetary. With computer-mediated judgments we move from information technologies to knowledge technologies to judgment technologies, making manifest the historical role of inscriptionality and its material and conceptual tools as technologies of faculties of mind and action. Increasingly, to a degree larger than now, computational technologies will be not only supplements for organizing and recommending judgments for us to make, but they will directly be making and enacting judgments and we will be inputs in their systems. And increasingly, we will feel comfortable with these judgments in ways that may not be true now. Technology, like custom, tends to seduce us all.

Departing from a traditional AI paradigm of symbol manipulations, where machines are programmed *a priori* to capture all the details of any given situation, "machine learning" derives values based on iteratively weighted input data. Much depends on the initial parameters for learning and for supervised learning, of course, but once a training set has been learned, then this learning can be extended to new instances, modifying the algorithmic weighting mechanisms themselves. These new instances allow machine learning to be both predictive and generative of new knowledge. Technologies that involve weighting are inherently judgmental, even if initially. As Ethem Alpaydin (2016) writes,

> Every learning algorithm makes a set of assumptions about the data to find a unique model, and this set of assumptions is called the *inductive bias* of the learning algorithm.
>
> This ability of generalization is the basic power of machine learning: it allows going beyond the training instances. Of course, there is no guarantee that a machine learning model generalizes correctly—it depends on how suitable the model is, and how well the model parameters are optimized—but if it does generalize well, we have a model that is much more than the data. A student who can solve only the exercises that the teacher previously solved in class has not fully mastered the subject: we want them to acquire a sufficiently general understanding from those examples so that they can also solve new questions about the same topic. (p. 42)

Machine learning must not just generalize, but it also needs to be able to adjust to deviant cases. It must be able to learn about particulars by their particular actions, not just by statistical averages. But through concentrating upon particulars in their histories, machine learning can also be susceptible to not only "catastrophic forgetting" (or "catastrophic inference") (French, 1999), but also "catastrophic remembering" (Sharkey & Sharkey, 1995) of cases, giving undo weight upon earlier learned events.[6]

The extension of artificial intelligence to sense in the lived world constitutes an important step in the attempt to model the particular as particular. Seeing particulars in terms of emergent dispositional powers opens up paths for better understanding cell expressions in cancer and other natural phenomena as dispositional expressions in environments of affordances and suppression, as well as social phenomena. However, to reemphasize Alpaydin's point above, what one ends up with still depends upon the empirical situation chosen, the inductive bias, the mediating parameters and other functions imposed and utilized, and lastly, the ends for which the processing occurs. In the social sciences, there is also the issue that the object of study and the method of studying it are shaped by the initial ontology chosen (e.g., in psychology whether one considers the brain or culture as "mind"). In the attempt to model particulars as particulars in their temporal expressions remains the problem of not only inscription, but also representation, in the choice of and application of data, in the processing of data, in the understanding of results, and in the social use of these results. Strong documentarity can come in the back door of weaker documentarity.

Conclusion

In this book, we have examined a broad range of different types, genres, and modes of inscription by which beings become evident and are taken as evidence. We have examined this from an aspect of strong documentarity by class categories and from the aspect of weaker documentarity due to the powers of entities to push back on their representations and to assert their own powers of presence. We have also seen, however, that such empiricism is inscribed within disciplinary assumptions and ontologies, parameters of measure, social networks, and practical ends.

The viewpoint of this book has been that of the philosophy and practice of evidence from the perspective of representation. This corresponds to documentation studies in my field, library and information science. In these studies, documents are usually taken as evidence of some content that they are said to represent.

I have suggested, however, that documentation studies can also be studied performatively, that is, in terms of what documents do, namely, a certain sense of "documentality." Representation is certainly an important part of what they do, but we can also study documents in terms of their performances secondary to representation.

For example, we have discussed Latour's conception of maps as indexes for navigating, Kohn's conception of forest sounds as indexes of affective events, avant-garde art and literature's view of representation as devices for epistemic and political ends, and we may also recall here Michael Buckland's discussion of passports as materials used for entering and exiting countries in his 2014 article, "Documentality Beyond Documents" (Buckland, 2014). There is a rich tradition of such works by authors within or touching upon documentation studies: the writings Buckland and of Bernd Frohmann in documentation studies, Johanna Drucker in information visualization and digital humanities, Maurizio Ferraris in philosophical approaches to documentality, and of course, that of others.

The representational imagination, however, is a powerful tool in thinking about evidence and entities, even when it ends (because it begins) in a world of fantasy. Especially in moments of crisis, we are all most likely Platonists, rather than Aristotelians: clutching to hopes of a medical cure when ill, looking for a permanent fix to political or institutional problems, seeking absolute truth in science, or asking God to save us at the end of our own or another's life. We seek a truth, instead of what can only be as true as it can be. However, this can lead to a very unhealthy politics of prejudice, a neglect of science and scholarship, and even a bibliographic mystification of the function of libraries as knowledge institutions (Day, 2019). On the other hand, beliefs in transcendental or "fictional" truths can also give us the hope and faith to get through the difficulties and fragility of our lives, to embark on cures for illnesses, fixes for political institutions, and to do research. It can also drive us in the necessary critical review of science, scholarship, and other practices and theory.

Representation, like all devices of inscription and their practices, requires us to think carefully about when to deploy it, because it has consequences not just in knowledge or documentation activities, but also in moral or "practical" activities (such as defining right-bearing entities and activities), and, of course, aesthetic judgments (such as judging what is "likeable" or not). It is sort of paradoxical to put it this way, but what becomes evident and what appears to us as evidence both depend on the mechanisms of inscription by which we represent what is present so that the present can be represented.

Notes

Introduction

1. I use the terms "Western metaphysical tradition," "Western culture," and like terms somewhat hesitantly because they suggest that such traditions designate permanently delineated national regions, civilizations, "peoples," and so forth, to which the beliefs and practices that I discuss in this book belong. What I mean by such grandiose terms, however, are cultural "family resemblances" of certain dominant beliefs and practices. I see metaphysical traditions as sort of long-running cultural psychologies, with all the problematic qualities of any cultural psychology, regarding ethnic and national bodies, problems of stereotypes, and so on.

2. Michael Buckland, in a recent paper on a 1948 paper by Briet's student, Robert Pagès, has suggested that such an investigation on this question may have been begun by Pagès with his notion of there being entities that are "auto-documents" (Buckland, 2017). As I will suggest in this book, something like the concept of an "auto-document" can have several different meanings, from being that of a semiotic or "social fact" (Ferraris, 2013) to that of being a natural power of self-expression. In Briet's (1951, 2016) notion of documentation, an animal must be captured and put in a knowledge or information structure in order for it to become evidence; for Briet, documents are, necessarily, social facts.

Chapter 1

1. Der Titel nennt den Versuch einer Besinnung, die im Fragen verharrt. Die Fragen sind Wege zu einer Antwort. Sie müßte, falls sie einmal gewährt würde, in einer Verwandlung des Denkens bestehen, nicht in einer Aussage über einen Sachverhalt.

    Der folgende Text gehört in einen größeren Zusammenhang. Es ist der seit 1930 immer wieder unternomme Versuch, die Fragestellung von "Sein und Zeit" anfänglicher zu gestalten. Dies bedeutet: den Ansatz der Frage in "Sein und Zeit" einer immanenten Kritik zu unterwerfen. Dadurch muß deutlich werden, inwiefern die kritische Frage, welches die Sache des Denkens sei, notwendig und ständig zum Denken gehört. Dem zufolge wird sich der Titel der Aufgabe "Sein und Zeit" ändern.

Wir fragen:

1. Inwiefern ist die Philosophie im gegenwärtigen Zeitalter in ihr Ende eingegangen?
2. Welche Aufgabe bleibt dem Denken am Ende der Philosophie vorbehalten?

2. Giorgio Agamben (2009) argues from a reading of Plato and others that paradigms are particulars compared to particulars—i.e., analogues—displacing a universal-particular framework. While this is an interesting conception, in the case of Plato's works it seems to me that though paradigms are spoken of as analogues, they are analogues that model a path forward in inquiry in such a manner that their exemplarity exceeds their particularity. Paradigms, in this sense, are exemplary by virtue of their representing some larger idea. They don't escape a universal-particular framework.

3. I discussed this in *The Modern Invention of Information: Discourse, History, and Power* (Day, 2001).

4. This "immanent" critique of *Being and Time* follows that book's own immanent critique of consciousness in Husserl's phenomenology. For more on the relation of *Being and Time* to Husserlian immanence, see Carmen's *Heidegger's Analytic: Interpretation, Discourse, and Authenticity in Being and Time*, pp. 86–93.

5. Paul Edwards has discussed data smoothing in scientific models of climate science (Edwards, 2010).

6. Bernd Frohmann, in his book *Deflating Information: From Science Studies to Documentation*, writes in regard to journal articles, for example: "The third consequence for the journal article follows from the first two. If neither truth nor epistemic significance are inherent properties of documents, then neither is information. Whether an article is informing depends on what happens to it later on, as it becomes implicated in particular epistemic alignments. Its informing character is therefore emergent, an effect of its enrollment in further projects, rather than a consequence of the completeness and presence of epistemic content. The article communicates no information by itself, whether information be conceived as inhering in the text or in 'interpretations' resident in the minds of representing subjects. Instead, its significance depends upon the temporal, open-ended, and in principle incomplete epistemic alignments in which its inscriptions are engaged" (Frohmann, 2004, p. 138).

7. Let us take a fictional example: Bill's *person* may indeed be said to have the character of "being" a jerk, if this is seen over time. "Jerkness," however, doesn't necessarily belong to Bill's *self* like a car belongs to Bill nor does it belong to him like his arm or brain belongs to him (as an innate part of his body), but rather, it is a moral quality given to observed behaviors that others see, and these have a stronger

correlation to innate dispositions of Bill's being if these behaviors are displayed over time, rather than in a few inconsiderate acts. The self is a set of dispositions, not a set of fixed transcendental traits. (Fortunately, "Bill," like the rest of us, can change his moral being, since it is made up of cultural expressions in social situations, and so inappropriate behaviors can be unlearned. Indeed, the task of moral improvement over a lifetime is what makes up "virtue ethics.") "Bill is a jerk," as a statement of dispositions, can only fairly be asserted by observing his behavior over time and seeing that such dispositional properties for Bill are not necessarily forever, no more than any of our own selves' expressive dispositions are. On the other hand, social understandings of "jerk behavior" may be quite common and accepted, so it may be right for us to assert in this case that, as a person, Bill's acts fit this profile (as ours do too, when we act in such ways as to conform to this moral category), and, in this sense, Bill really is a jerk. People are not transcendentally good or bad, but rather, they act in good and bad manners, and from that we infer that they "have" these moral qualities. (But, as Wittgenstein warned us, we have to be careful with this grammar of "to have," because it not the same in all our uses of the term.) The same is true of "having knowledge," "having intelligence," and so on; mental dispositions and expressions can only be *correlated* through seen or measured expressions.

Chapter 2

1. See also Alice L. Conklin's *In the Museum of Man: Race, Anthropology, and Empire in France, 1850–1950* (Conklin, 2013).

2. It may be possible, however, to see a difference in these two modes of representation, similar to making a distinction between scientific documentaries and fictional documentaries in film (or beginning historically earlier, the difference between the emergence of social science claims and fictional realism—this latter which we will discuss in a later chapter). The difference involves a problem that is central to this book, namely, the role of representation in knowledge processes. In the former, representations are inserted into discursive arguments; so, for example, the renowned anthropologist Napoleon Chagnon's documentary films of the Yanomami indigenous tribe might be used as evidence within claims of their ferocity, among other descriptive claims. This parallels the use of natural science experiments as evidence within claims about the nature of physical reality; there are resistances and supports to this evidence outside of the content of the represented evidence itself. This is very different from the case of the 1922 film, *Nanook of the North*, or earlier in Balzac or Flaubert's depictions of French society in the first half of the nineteenth century, where the representations themselves claim to contain the evidence of what is. The role of evidence in the first is indexical in regard to not only the presenter's argument, but also in regard to the evidence and arguments of others. The purpose of scholarly or scientific works are to put something into discursive play—they put into discursive play an argument for what is. In the second case, the representation

is iconic; in fact, it aims to erase its mode of production so that it produces a totally aesthetic representation.

3. On this latter, see Bertrand Russell's introduction to Wittgenstein's *Tractatus Logico-Philosophicus*.

Chapter 3

1. Scholastic theological arguments, on the other hand, seem to lack this temporal and experiential horizon, since truth appears from out of argumentative, logical-deductive modes of reasoning rather than descriptive, empirical ones. Rationalist devices are used to keep understanding "pure" and unpolluted by experience. The experiential horizon disappears, and with this, the particulars are analytically derived from universals. History, here, appears only as an explanation of rational arguments; scholastic arguments are distant from phenomenological perspectives, so they are language games that are accessible only to the philosophically initiated. The point of view with these is that of viewing God as a rational being of beings, whose phenomenological manifestations are inflected in the confused human world, like a stick that appears bent by the water it sits in. Scholastic argument attempted to discover the true rationality beyond the irrationality of the phenomenological world.

Chapter 4

1. Long ago, in an interview, I said that "we must now all be information professionals" (Day & Pyati, 2005). Now I would add, "We must also all be literary critics and humanists," since so much of our lives are mediated by literary forms and rhetorical devices through information devices.

2. Flaubert's novel demonstrates that in bourgeois society the other person is taken as a personification of poor moral qualities, but the self is understood as a limitless potentiality of both good and bad actions. This is a particular social extension of the tendency in modern psychology to see others as person types, with known causalities, and so fixed intentionality, while one's self remains a toolbox of potentialities of choices, actions, and in short, "freedom" The formal irony in the novel, and so many other novels since, is that characters see one another as being characters, but one's self as not being a character, and finally from such a viewpoint—particularly those written in the first person—that the story is not a story but a depiction of reality.

3. These and other such rural values in a Chinese context can be found in Eli Blevis and Shunying An Blevis's account (Blevis, 2018).

4. "If it were the intention of the press to have the reader assimilate the information it supplies as part of his own experience, it would not achieve its purpose. But its intention is just the opposite, and it is achieved: to isolate what happens from

the realm in which it could affect the experience of the reader" (Benjamin, 1968a). Conversely, in modernity the mass media also commonly expands a microscopic event so that it seems the whole of the viewer's reality. In so doing, it can give undue weight to issues that were previously absent or minor and distort the viewer's or reader's perception into being a product of an increasingly extreme attention economy.

Chapter 5

1. An earlier exploration of this section on jokes was "Rethinking unsaid information: jokes and ideology" (Day & Ma, 2011).

2. This and the following section of the chapter were stimulated by Professor Peggy McCraken's paper, "Metamorphosis and Living Death," which she gave at Indiana University at Bloomington in the autumn of 2016. I had long been wanting to discuss the absence of trauma in discourses on information, and Professor McCraken's discussion of the Old French *Philomena*, a twelfth-century translation of the Philomela story from Ovid's *Metamorphoses*, particularly as it intersected in time with a lecture I attended by Professor Cindy Bethel on her study of a robotic dog, Therabot, as a device for eliciting narratives from adult and child trauma victims, provided an impetus to discuss these as modes of evidence in the present book. I am grateful to both these researchers for their presentations at Indiana University and to Professor McCraken for sharing with me a copy of her text to read and her kind correspondence afterwards. I am also grateful to my colleague at Indiana University, Professor Selma Šabanović, for organizing and inviting me to Professor Bethel's lecture.

3. For a fuller account of this term, particularly in French psychoanalysis, see House (2015).

Chapter 6

1. I am grateful to my Brazilian colleague Professor Lídia Freitas for introducing me to the case of the SPI.

2. http://www.oabpa.org.br/index.php/25-noticias/4573-cfoab-comissao-da-verdade-da-escravidao-negra-toma-posse-na-oab-nacional.

3. Tool use is sometimes used as a distinguishing trait of human beings, but this seems to me problematic, except in regard to what was just discussed. It is often assumed that only recently have human beings noticed that nonhuman animals use technology (e.g., crows with sticks). However, it seems to me that the skills and abilities of animals to negotiate their environments by means of building nests and so forth have always been observed and noted. The difference that has been asserted is that of the ability to *represent or imagine* these activities and to transfer them to other activities. The abstraction of immediately afforded *techne* to other activities is

key to understanding the difference between *techne* and technology in Heidegger's works, for example. It is *technological transfer* that constitutes the core of modern technology and human reason in the philosophical tradition, not "tools" as immediate affordances. The role of representation or "imagination" in this regard is what is of importance.

4. In my previous book, *Indexing It All*, I discussed androids in terms of documentary identity. Related to this, it might be mentioned that androids could be subject to a similar analysis as Heidegger's analysis of animals in terms of domesticity. The phenomenon of the "uncanny valley" is perhaps a poor criterion for analyzing androids, for the very category itself is rather uncanny for its lack of specific meaning. An analysis of androids in terms of their inability "to go along with us" for very long at all perhaps can better express the lack of human *pathos* toward androids, at least at this stage in their technological development.

5. The difficulty of some of these positioning narratives for rights and identity can be seen in the case of Kohn's (2013) work, which raises the perplexing problem of how it is that people who supposedly treat their environment and its beings as fellow subjects can also treat those other subjects as if they were mere objects—killing, and in the case of nonhuman animals, devouring them, with seemingly little hesitation.

Chapter 7

1. The mid-twentieth-century French philosopher of phenomenology and aesthetics, Raymond Bayer (professor of philosophy at the Sorbonne, who is cited in Briet's *Qu'est-ce que la documentation?*), gave the opening lecture in the second year of the L'Union française des organismes de documentation's Cours Techniques de Documentation (1946–1947) (an educational course that would later evolve into the curriculum of Briet's l'Institut national des techniques de la documentation). In this lecture, Bayer suggested that documents are the forms imposed upon floods of information during modernity. He viewed documentary technique lying within the hands of documentary specialists who could control information, rather than be controlled by it. Nonetheless, his remarks deserve our reflection today, as every person sees him- or herself as Bayer's "man of science" through the Internet, which, as we have discussed in this book, mediates our worlds as second-hand knowledge. Addressing an audience of documentalists, Bayer opens his lecture with this theme of information as being the producer and product of the human mind or spirit (*l'esprit*—I will keep the French term in the translation below, as Bayer plays with this dual meaning in his sorcerer and his apprentice metaphor and also both meanings play back and forth with one another throughout his lecture). Documentary techniques and technologies (today in the form of algorithms) as tools for the organization, but also as the producers, of "information," assume the role of Kantian apperception—or mind—in constructing the world we live in. This is the

Notes

philosophy of what I called in *Indexing It All* the "modern documentary tradition"; the idea of documentary technology as the driving force of the spirit of existence in modernity (the central theme of Briet's 1951 book). Documentary technologies and their products are not just categories of organizing the world of information, but are productive of that world, as a lived experience. This is the essence of "the information age." Bayer writes (in translation):

> You [documentalists] are the group of specialists that prevents the Sorcerer from remaining apprentice. In modern times, this has been the adventure of the man of science; he has, by the spirit [*l'esprit*: mind or spirit], unleashed *l'esprit*: the producer has submerged himself under his products. Information leads us, it knocks on our door, it hits our private thoughts, it gnaws at our minutes of meditation, it upsets our inner life. The actuality and the passing moment ring and resound near us like an intrusive telephone. It organizes chaos. We are, according to the mood, the slaves or the clients of this imperious master. . . . Because documentation is the form given to all information. It is the information that is craved and grasped by the *l'esprit*, bearing and keeping all the intentions of the *l'esprit*. (Bayer, 1946–1947)

I am grateful to Professor Claire Scopsi for her assistance in locating this material.

2. I was a middle and high school librarian at the time of the appearance of the graphic user interface Internet, and so I literally witnessed the massive growth of the Internet week by week with such searches: one week several thousand hits for a search term, the next week or two, several million hits for the same term, and so forth. School librarians at the time first thought that they could create user guides for their students of the best sites, but this very quickly became comical as the number of sites massively grew week by week.

3. So, for example, if one looks in Google for "hot dog," the highest ranked results are for what in English is also called a "frankfurter." To find a dog that is hot on a summer's day, one has to modify the search in such a way as to get around the popular frankfurter. More than any theoretical argument, Google Search shows the superiority of social epistemology over professionally imposed categories for establishing user relevancy. It is a wonderful example of the importance of Blair's insight that earlier IR was misled by a philosophy of language based on the notion that meaning was created by categories rather than by language use.

4. An astute reader of the manuscript of this book argued that there is nothing to stop a professional cataloger from using the formulation "dogcat" when describing an object. He or she argued that LCSH itself is not a barrier to this, since it is simply a set of terms. The larger point was that there's no need to assert a notion of post-documentation technologies, since traditional and newer documentation techniques and technologies both follow rule-based practices. In this sense, they are both Wittgensteinian language games.

My response is that while it is true that both natural language and professional cataloging are rule-based activities, the nature of the rules is very different. LCSH is a controlled vocabulary within a professional practice of naming items according to that language, nothing else. The terms within it constitute identities marked by differences within a set number of terms. Natural language functions very differently in that there are wide varieties of synonyms, multiple senses to terms, and other sources of ambiguity. Meaning comes about in natural language through use, not through identity and differences within a "closed vocabulary" structure. The professional cataloger can, of course, use any term that he or she wants, and sometimes does this in error. But the whole point of a controlled vocabulary is to try and stabilize the relation between a term and its concept by using an authorized vocabulary. An entry such as "dogcat" in LCSH would be seen as an error, whereas its use in user-centered tagging, for example, would more likely be seen as simply a user's choice of terms, however idiosyncratic or useful or not.

There are many means of algorithmically mediated searching, with or without user prompting, that attempt to substitute for controlled vocabulary toward gaining greater precision or relevancy, such as keywords in context searching, keywords out of context, and of course, link analysis such as PageRank, or even social or geographical context mediation. All these techniques attempt to "control" natural language in ways different than through traditional documentation techniques, but the goal is the same: to create conceptual reference. However, newer vocabularies can incorporate a broader sense of terms as used in natural language usage by following vectors of word use, where professional controlled vocabularies such as LCSH incorporate very limited senses of a term in order to keep very defined conceptual referents.

Traditional controlled vocabulary is characterized by minimal grammatical sense and a controlled sense of reference by means of strict differences between terms with little ambiguity being allowed. Natural language is characterized in most cases by high degrees of ambiguity that are resolvable by grammatical context and real-world use. While it is true that libraries, for example, are "real-world" sites, the use of vocabulary there, at least in the case of subject searching using LCSH, is so peculiar as to be a barrier to most searchers. There are few, if any, "real-world" examples outside of libraries for using LCSH formulations, such as saying, "Midway, Battle of, 1942."

5. See, for example, Jan Pieter Barbian's account of the Nazi usurping of the publishing industries and the library committees and institutions in *The Politics of Literature in Nazi Germany: Books in the Media Dictatorship* (Barbian, 2013).

6. I am grateful to my colleague at Indiana University, David B. Leake, for these references.

References

Ades, D., Baker, S., & Bradley, F. (2006). *Undercover surrealism: Georges Bataille and Documents*. Cambridge, MA: MIT Press.

Agamben, G. (2009). *The signature of all things: On method*. New York: Zone Books.

Alpaydin, E. (2016). *Machine learning*. Cambridge, MA: MIT Press.

Auerbach, E. (2003). *Mimesis: The representation of reality in Western literature*. Princeton, NJ: Princeton University Press.

Balnaves, M., & Willson, M. (2011). *A new theory of information and the Internet: Public sphere meets Protocol*. Digital Formations. New York: Peter Lang.

Balzac, H., & Saintsbury, G. (1901). *The works of Honoré de Balzac; With introductions by George Saintsbury* (Vol. 1). Boston, MA: Dana Estes.

Barbian, J.-P. (2013). *The politics of literature in Nazi Germany: Books in the media dictatorship*. London: Bloomsbury Academic.

Bataille, G. (1985). *Visions of excess: Selected writings, 1927–1939*. Minneapolis: University of Minnesota Press.

Bayer, R. (1946–1947). *Lecon 1, Chapitre 1: Documentation et Documentologie: Documentation et philosophie*. Paris: Cours techniques de documentation.

Benjamin, W. (1968a). On some motifs in Baudelaire. *Illuminations* (pp. 155–200). New York: Schocken.

Benjamin, W. (1968b). The work of art in the age of mechanical reproduction (H. Zohn, Trans.). In H. Arendt (Ed.), *Illuminations* (pp. 217–251). New York: Schocken.

Bishop, C. A. (2012). *Access to information as a human right*. El Paso, TX: LFB Scholarly Publishing.

Blair, D. C. (2006). *Wittgenstein, language, and information: Back to the rough ground!* Dordrecht: Springer.

Blevis, E., & Blevis S. A. (2018). Design inspirations from the wisdom of years. Paper presented at the 2018 Designing Interactive Systems Conference, Hong Kong, China.

Börner, K. (2015). *Atlas of knowledge: Anyone can map*. Cambridge, MA: MIT Press.

Briet, S. (1951). *Qu'est-ce que la documentation?* Paris: Éditions documentaires, industrielles et techniques.

Briet, S. (1954). Bibliothécaires et documentalistes. *Revue de documentation, 21*, 41–45.

Briet, S. (2006). *What is documentation? English translation of the classic French text*. (R. E. Day, L. Martinet, and H. Anghelescu, Eds.). Lanham, MD: Scarecrow Press. Originally published in French in 1951 as *Qu'est-ce que la documentation?* (Paris: Éditions documentaires, industrielles et techniques).

Broussard, R., & Doty, P. (2016). Toward an understanding of fiction and information behavior. *Proceedings of the Association for Information Science and Technology, 53*, 1–10. doi:10.1002/pra2.2016.14505301066.

Buckland, M. K. (2014). Documentality beyond documents. *Monist, 97*(2), 179–186. https://escholarship.org/uc/item/55v7p74x.

Buckland, M. K. (2017). Before the antelope: Robert Pagès on documents. Paper presented at the Document Academy Conference 2017. https://ideaexchange.uakron.edu/docam/vol4/iss2/6/.

Burrell, J. (2016). How the machine "thinks": Understanding opacity in machine learning algorithms. *Big Data & Society, 3*(1), 1–12.

Calarco, M. (2015). *Thinking through animals: Identity, difference, indistinction*. Stanford, CA: Stanford University Press.

Canêdo, F. (2013, April 19). Documento que registra extermínio de índios é resgatado após décadas desaparecido. *Estado de Minas*. http://www.em.com.br/app/noticia/politica/2013/04/19/interna_politica,373440/documento-que-registra-exterminio-de-indios-e-resgatado-apos-decadas-desaparecido.shtml.

Céline, L.-F. (2006). *Journey to the end of the night*. (William T. Vollman, Ed.; Ralph Manheim, Trans.). New York: New Directions.

Conklin, A. L. (2013). *In the museum of man race, anthropology, and empire in France, 1850–1950*. Ithaca, NY: Cornell University Press.

Coolidge, C. (1974). *The maintains*. San Francisco, CA: This Press.

Day, R. E. (2000). The "Conduit Metaphor" and the nature and politics of information studies. *Journal of the Association for Information Science and Technology, 5*(1), 805–811.

References

Day, R. E. (2001). *The modern invention of information: Discourse, history, and power.* Carbondale: Southern Illinois University Press.

Day, R. E. (2006). "A necessity of our time": Documentation as "cultural technique." In S. Briet, *What is documentation? English translation of the classic French text* (pp. 46–63). Lanham, MD: Scarecrow Press.

Day, R. E. (2014). *Indexing it all: The subject in the age of documentation, information, and data.* Cambridge, MA: MIT Press.

Day, R. E. (2019). Right-wing populism, information, and knowledge. *Logeion: Filosofia da Informação, 5*(2), 38–54.

Day, R. E., & Pyati, A. K. (2005). "We must now all be information professionals": An interview with Ron Day. *InterActions: UCLA Journal of Education and Information Studies, 1*(2). https://escholarship.org/uc/item/6vm6s0cv#author.

Debaene, V. (2014). *Far afield: French anthropology between science and literature.* Chicago: University of Chicago Press.

DeJean, J. (1991). *Tender geographies: Women and the origins of the novel in France.* NY: Columbia University Press.

Deleuze, G. (2012). *Gilles Deleuze from A to Z./Interviewer: C. Parnet.* Cambridge, MA: Semiotext(e).

Deleuze, G., & Guattari, F. (1987). *A thousand plateaus* (B. Massumi, Trans.). Minneapolis: University of Minnesota Press.

Derrida, J., & Bennington, G. (2009). *The beast and the sovereign.* Chicago: University of Chicago Press.

Doty, P., & Broussard, R. (2017). Fiction as informative and its implications for information science theory. In S. Erdelez & N. K. Agarwal (Eds.), *Proceedings of the Association for Information Science and Technology* (pp. 61–70). Hoboken, NJ: Wiley. https://doi.org/10.1002/pra2.2017.14505401008.

Drucker, J. (2009). *SpecLab: Digital aesthetics and projects in speculative computing.* Chicago: University of Chicago Press.

Drucker, J. (2014). *Graphesis: Visual forms of knowledge production.* Cambridge, MA: Harvard University Press.

Duckworth, D., Henkel, Z., Wuisan, S., Cogley, B., Collins, C., & Bethel, C. (2015). Therabot: The initial design of a robotic therapy support system. Paper presented at the Tenth Annual ACM/IEEE International Conference on Human-Robot Interaction Extended Abstracts, Portland, Oregon, USA.

Edwards, P. N. (2010). *A vast machine: Computer models, climate data, and the politics of global warming.* Cambridge, MA: MIT Press.

Ferraris, M. (2013). *Documentality: Why it is necessary to leave traces.* New York: Fordham University Press.

Flaubert, G., Brunetière, F., Arnot, R., Flaubert, G., Flaubert, G., & Flaubert, G. (1904). *Madame Bovary: A tale of provincial life* (Vol. 1). Chicago: Magee.

Flint, C. G., I. Kunze, A. Muhar, Y. Yoshida, & M. Penker. (2013). Exploring empirical typologies of human-nature relationships and linkages to the ecosystem services concept. *Landscape & Urban Planning, 120,* 208–217.

Fonseca, M. O. (1999). Informação e direitos humanos: Acesso às informações arquivísticas. *Ciência da Informação, 28*(2). 146–154.

Foucault, M. (1971). *The order of things: An archaeology of the human sciences.* New York: Pantheon Books.

French, R. M. (1999). Catastrophic forgetting in connectionist networks. *Trends in Cognitive Sciences, 3*(4), 128–135.

Freud, S. (1961). *Beyond the pleasure principle (The standard edition).* (James Strachey, Trans.). New York: Liveright Publishing.

Freud, S. (1980). *The interpretation of dreams.* New York: Avon.

Frohmann, B. (2004). *Deflating information: From science studies to documentation.* Toronto: University of Toronto Press.

Freud, S. (1989). *Jokes and their relation to the unconscious.* New York: W. W. Norton.

Harré, R. (1989). The "self" as a theoretical concept. In M. Krausz (Ed.), *Relativism: interpretation and confrontation* (pp. 387–417). Notre Dame, IN: University of Notre Dame Press.

Harré, R. (1995). Realism and an ontology of powerful particulars. *International Studies in the Philosophy of Science, 9*(3), 285–300.

Harré, R. (2002). Material objects in social worlds. *Theory, Culture, & Society, 19*(5–6), 23–33.

Harré, R., & Llored, J.-P. (2018). Procedures, products and pictures. *Philosophy, 93,* 167–186.

Harryman, C. (1989). *Animal instincts.* Berkeley, CA: This Press.

Harryman, C. (1992). *In the mode of.* Tenerife, Spain: Zasterle.

Harryman, C. (2001). *Gardener of stars.* Berkeley, CA: Atelos Press.

Harryman, C. (2008). *Adorno's noise.* Ithaca, NY: Essay Press.

Heidegger, M. (1971). The way to language. In *On the Way to Language* (Peter D. Hertz, Trans., pp. 111–136). New York: Harper & Row.

References

Heidegger, M. (1977a). The age of the world picture. In D. F. Krell (Ed.), *Martin Heidegger: Basic writings from Being and Time (1927) to the Task of Thinking (1964)* (pp. 115–154). New York: Harper & Row.

Heidegger, M. (1977b). The end of philosophy and the task of thinking. In D. F. Krell (Ed.), *Martin Heidegger: Basic writings from Being and Time (1927) to the Task of Thinking (1964)* (pp. 373–392). New York: Harper & Row.

Heidegger, M. (1977c). The question concerning technology. In *The question concerning technology and other essays* (William Lovitt, Trans., pp. 3–35). New York: Harper.

Heidegger, M. (1979). *Nietzsche: The will to power as art* (Vol. 1). San Francisco: Harper & Row.

Heidegger, M. (1995). *The fundamental concepts of metaphysics: World, finitude, solitude*. Bloomington: Indiana University Press.

House, J. (2015). In French psychoanalysis: The long life of Nachträglichkeit. The first hundred years, 1893 to 1993. *Psychoanalytic Review, 102*(5), 683–708.

Kant, I. (2000). *Critique of the power of judgment* (P. Guyer, Trans.). Cambridge: Cambridge University Press.

Kant, I. (2013). *An answer to the question: "What is enlightenment?"* London: Penguin Books.

Kohn, E. (2013). *How forests think: Toward an anthropology beyond the human*. Berkeley, CA: University of California Press.

Latour, B. (1996). Ces réseaux que la raison ignore—laboratoires, bibliothèques, collections. In Marc Baratin & Christian Jacob (Eds.), *Le pouvoir des bibliothèques: La mémoire des livres dans la culture occidentale*. (pp. 23–46). Paris: Albin Michel.

Lhermitte, J.-F. (2015). *L'animal vertueux dans la philosophie antique à l'époque impériale*. Paris: Classiques Garnier.

Martincorena, I., Fowler, J. C., Wabik, A., Lawson, A. R. J., Abascal, F., et al. (2018). Somatic mutant clones colonize the human esophagus with age. *Science, 362*(6417), 911–917.

Mathiesen, K. (2015). Human rights as a topic and guide for LIS Research and Practice. *Journal of the Association for Information Science and Technology, 66*(7), 1305–1322.

McNeill, W. (1999). Life beyond the organism: Animal being in Heidegger's Freiburg Lectures, 1929–30. In H. P. Steeves (Ed.), *Animal others: On ethics, ontology, and animal life* (pp. 197–248). Albany, NY: SUNY Press.

Medina, E., & Wiener, I. S. (2016). Science and harm in human rights cases: preventing the revictimization of families of the disappeard. *Yale Law Journal Forum, 125*, 331–342.

Mignolo, W. (2003). *The darker side of the Renaissance: Literacy, territoriality, and colonization* (2nd ed.). Ann Arbor: University of Michigan Press.

Mignolo, W. (2006). Citizenship, knowledge, and the limits of humanity. *American Literary History, 18*(2), 312–331.

Mignolo, W. (2011). *The darker side of Western modernity: Global futures, decolonial options.* Durham: Duke University Press.

Naqvi, Y. (2006). The right to truth in international law: Fact or fiction? *International Review of the Red Cross, 88*(862). https://www.icrc.org/en/international-review/article/right-truth-international-law-fact-or-fiction.

Otlet, P. (1934). *Traité de documentation: Le livre sur le livre: Théorie et pratique.* Brussels: Editiones Mundaneum, Palais Mondial.

Pelizzon, A., & Gagliano, M. (2015). The sentience of plants: Animal rights and rights of nature intersecting? *Australian Animal Protection Law Journal, 11*, 5–11.

Reddy, M. J. (1979). The conduit metaphor: A case of frame conflict in our language about language. In A. Ortony (Ed.), *Metaphor and thought* (pp. 284–310). Cambridge: Cambridge University Press.

Rieder, B. (2012). What is in PageRank? A historical and conceptual investigation of a recursive status index. *Computational Culture, 2.* http://computationalculture.net/what_is_in_pagerank/.

Safian, L. A. (1981). *The giant book of insults.* Secaucus, NJ: Castle Books.

Sharkey, N., & Sharkey, A. (1995). An analysis of catastrophic interference. *Connection Science, 7*, 301–329.

Sorabji, R. (1993). *Animal minds and human morals: The origins of the Western debate.* Ithaca, NY: Cornell University Press.

Stone, C. D. (2010). *Should trees have standing? Law, morality, and the environment* (3rd ed.). New York: Oxford University Press.

Thomas, N. (2012). The algorithmic representation of need. *Culture Digitally*, June 24. http://culturedigitally.org/2012/06/the-algorithmic-representation-of-need/.

Thomas, N. (2018). *Becoming-social in a networked age.* New York: Routledge.

Tort, P. (2001). *Darwin and the science of evolution.* New York: Harry N. Abrams.

Toupin, L. (2018). *Wages for housework: A history of an international feminist movement, 1972–1977.* Vancouver and London: UBC Press and Pluto Press.

Walsh, J. A. (2012). "Images of God and friends of God": The holy icon as document. *Journal of the American Society for Information Science and Technology, 63*(1), 185–194.

References

Watten, B. (1984). *Total syntax*. Carbondale: Southern Illinois University Press.

Watten, B. (1985). *Progress*. New York: Roof Books.

Watten, B. (1988). *Conduit*. Berkeley, CA: GAZ.

Watten, B. (2003). *The constructivist moment: From material text to cultural poetics*. Middletown, CT: Wesleyan University Press.

Watten, B. (2016). *Questions of poetics: Language writing and consequences*. Iowa City: University of Iowa Press.

Watts, J., & Rocha, J. (2013, May 29). Brazil's "lost report" into genocide surfaces after 40 years. *Guardian*. https://www.theguardian.com/world/2013/may/29/brazil-figueiredo-genocide-report.

Wiener, N. (1961). *Cybernetics: Or control and communication in the animal and the machine* (2nd ed.). Cambridge, MA: MIT Press.

Wilson, P. (1983). *Second-hand knowledge: An inquiry into cognitive authority*. Westport, CT: Greenwood Press.

Index

"Aboutness," 42, 128, 129, 141–143, 148
Actor network theory, 22, 23, 30
Aesthetic/experiential index of documentary representation, 40
Aesthetic judgments, 149, 152
Aesthetics ("feelings"). *See also* Bataille, Georges; Feelings
  documentarity and, 1, 2
  information and, 2–3
  Kantian, 86, 87
Affect(s), 23. *See also* Feelings
  directionality of, 24–25
  documentarity and, 20
  entities and, 20, 24–25
  Heidegger and, 20
  Kohn and, 128, 129
  Latour and, 23, 24
  literature and, 67, 69, 85, 86
  post-documentation and, 138
  self and, 127, 128
  semiotics and, 128, 129
  "sense" and, 20, 24–25
Affordances, 30–35
  dispositional powers and, 9, 31, 34
  dispositions and, 14–15, 21, 31–35, 111, 129, 137
  Heidegger and, 14–15, 18, 32
  Latour and, 8, 21, 22, 25, 27
  science and, 32, 35, 137
  types of, 21, 27, 32, 33, 111
After-affects (*Nachträglichkeit*), 104, 105

Agamben, Giorgio, 154n2
Agency rights, 111–112, 119, 131, 137. *See also* Expression: rights of
Aition (causes/causal affordances), 18, 32
Algorithms, 137, 140–142, 144, 145. *See also* Learning algorithms
Allegory, 57–59, 106–108
Alpaydin, Ethem, 149–150
Amazonian Kichwas. *See* Runa people in Ecuadorian Amazon
Androids, 158n4
Animal bodies. *See* Bodies
"Animal lovers," 124, 132
Animal rights, 123–126, 158n5
  conflicting with one another, 132
  rights of nature and, 118–119, 125, 130–134
  Kohn on, 132, 133, 158n5
Animals, 71, 106, 110, 119–121
  Aristotle on, 120, 121
  Balzac on, 70, 71
  and the big toe, 47, 48
  Briet on, 61–63, 110, 129, 153n2
  Buffon on, 71–72
  cybernetics and communication in, 16
  as documents, 61
  domestication, domesticity, and, 122–124, 130–132, 134, 158n4
  empathy and, 122, 126, 132, 133
  essence and, 3, 123, 129

Animals (cont.)
  habits, 72
  Heidegger, the world, and, 121–126
  humans compared with, 47, 70–72, 119–123, 126, 157n3
  indexicality and, 126, 129
  Kohn on, 128–129
  language and, 119, 121
  Lhermitte and, 120–122
  ontology and, 62, 120, 123–126, 129–131, 134
  representation and, 119–121, 157n3
  semiotics and, 126, 129, 133
  sensibility/perception, 119–122
  taxonomies and, 28–29, 62, 125, 130, 134
  transposability and, 122–123
Anthropologists, 23, 39
Anthropology, 37–39, 118, 126. *See also under* Latour, Bruno
  Bataille and, 38, 47–49
  Debaene on, 38, 39
  French, 39, 49
Anthropomorphism, 15, 33, 129, 132
Aristocracy (class), 73, 78, 80. *See also* "High style" literature
  in *Madame Bovary*, 73, 80, 81, 84
Aristotle, 120–122
Art. *See also* Icons and iconography
  function of, 57
Artificial intelligence, 134, 150
"As-if" structures, 146
  in fairy tales, folktales, and fables, 99, 106–108
Auerbach, Erich, 8, 52–56, 66–67
Austen, Jane, 81–82
Auto-documents, 153n2
Avant-garde, 46, 91. *See also* Bataille, Georges
  constructivism and, 89
  modern, 46, 66, 67, 85–89
  realism and, 66, 85–86, 96–97

  representation and, 85–87, 96–97, 151
  theory of the, 85–86

Ball, Hugo, 87
Balnaves, M., 40–41
Balzac, Honoré de, 70–72
Bataille, Georges, 45–49, 51
  anthropology and, 38, 47–49
  documentarity and, 48–49
  documents, documentation, and, 38, 40, 46
  Documents (magazine) and, 47
  epistemology and, 38, 47, 51
  ethnology and, 38
  literary works, 46–49, 51
  Otlet and, 38, 40, 45–49, 51
  and the philosophy of base materialism, 38, 45–47, 49, 51
Bayer, Raymond, 158n1, 159n1
Beauty. *See* Aesthetics; Bataille, Georges
Becoming, 51, 53
Being. *See also Dasein*; Existence
  modes of, 51, 64, 85, 126
  nature of, 138
Being and Time (Heidegger), 11–14, 19–20, 122, 154n4
Being-with (*Mitsein*), 122, 123, 126
Benjamin, Walter, 93–94, 147–148
Bethel, Cindy, 157n2
Biblical narratives, 54, 55, 60
Bibliographic positivism of Paul Otlet, 38, 40–45, 61
Biological affordances. *See* Physical/biological affordances
Bishop, C. A., 113
Blair, David C., 140
Bodies, 32, 33, 134
  legal standing of, 132–134
  rights of, 125, 130–133
Bourgeoisie, 65, 72, 77–80, 84, 115, 156n2. *See also Madame Bovary*

Index

in *Madame Bovary*, 70, 72–74, 78–81, 84, 115, 156n2
Brazil
   Indian Protection Service (SPI), 116, 117
   State Commission on the Truth of Black Slavery in Brazil, 118
Briet, Suzanne, 45, 52, 57, 153n2
   on animals/animal entities, 61–63, 110, 129, 153n2
   on *indice* (indexical sign), 56, 60–61
   ontology and, 57, 61–64, 119, 129
   science and, 63, 64
   What Is Documentation?, 3–5, 60–64, 125, 129, 159n1
Broussard, R., 84
Buckland, Michael, 151, 153n2
Buffon, Comte de (Georges-Louis Leclerc), 71–72

Calarco, M., 124
Categories, 98. *See also specific topics*
Category names. *See* Prepositions
Causes, 32. *See also* Aition
Céline, Louis-Ferdinand, 103–104
Center of calculation (*centre de calcul*), 27–29, 39, 63, 117
Chagnon, Napoleon, 155n2
Chains of reference, 26, 27
Character, moral, 69–70, 74, 154n7
   defined, 69
Christianity, 54–60
Classes, 25–26. *See also* Social class
Classification. *See* Ontology: typology/taxonomy/classification and; Taxonomy(ies); Typology
Colonialism, 116–117
"Coloniality of knowledge," 116
"Computer centers," 28, 119. *See also* Center of calculation
Computer-mediated judgments, 137–138, 148–149

Concepts, 57
Conceptualism, 88
Condensation of signification (Freud), 101
Conduit metaphor, 92
Conduit (Watten), 92–93
Constructivism, 22, 67, 98
   formalism and, 85, 89, 90
   literature and, 85, 86, 89, 90
   realism and, 67, 89–90
   Russian/Soviet, 86, 89
Containment, 96
   of trauma, 104–105
"Contextual" affordances, 31, 111
Controlled vocabularies, 140–143, 160n4
Coolidge, Clark, 88–89
Correia, Jader de Figueiredo, 116
Cultural affordances, 7, 21, 32. *See also* Affordances
"Cultural metaphysics," 14
Cultural technique(s), 62
   documentation as a, 62, 63
Cybernetics, 16–17, 132, 133, 139, 144

Dasein, 122–125, 130
   Heidegger on, 122, 123, 125
Data centers, 28, 119. *See also* Center of calculation
   libraries and other, 27–30 (*see also* Libraries)
Debaene, Vincent, 38–39
Decolonization, 116–117
Deleuze, Gilles, 24, 123–124, 130
Derrida, Jacques, 107–108
Dias, Marcelo, 118
Digital humanities, 97
Digital information infrastructure, 138
Displacement of signification (Freud), 101
Dispositional-affordance theory of expression, 31. *See also* Affordances

Dispositional powers
  affordances and, 9, 31, 34
  animals and, 110, 133
  entities and, 4, 31, 32, 110
  Harré on, 9
  innate, 2, 18, 22, 30
  particulars and, 2, 150
  rights and, 133
  theory of, 34
  weak documentary and, 2
Dispositional theories, 30. *See also* Dispositional powers: theory of
Dispositions, 137, 155n7
  affordances and, 14–15, 21, 31–35, 111, 129, 137
  entities and, 9, 14–15, 31, 32, 35, 111
  expression of, 30, 130, 137
  expressions and, 31, 33–35
  innate, 9, 21, 32
  ontology and, 30, 35
  physical, 33
  rights and, 111, 130, 133
  science and, 21, 32–33, 137
  and the self, 137, 155n7
Documentality: Why It Is Necessary to Leave Traces (Ferraris), 8–9
"Documentality beyond Documents" (Buckland), 151
Documentality vs. documentarity, 8–9
Documentarity. *See also specific topics*
  definition and nature of, 1, 116
  vs. documentality, 8–9
  epistemology and, 1, 2, 26, 37
  hallmark of documentarity in modernity, 74 (*see also* Modernity)
  metaphysics and, 1, 20, 38, 49, 68, 116–119
  modes of, 37, 85, 121 (*see also* Strong documentarity)
  ontology and, 1, 20, 21, 26, 67, 118
  origin of the term, 8
  science and, 39, 49, 56, 57
  strong form of (*see* Strong documentarity)

  weak form of (*see* Weak documentarity)
Documentary knowledge, 77, 109, 148. *See also* Secondhand knowledge
  science as, 40
Documentation
  literature and, 65, 68
  nature of, 61–63, 143
  Otlet on, 42–43
  science and, 49, 52, 53, 56, 60–64
  What Is Documentation? (Briet), 3–5, 60–64, 125, 129, 159n1
Documentation theory, 37, 52
  indexical sign (*indice*) and, 60–64
Documents, 142. *See also specific topics*
  definitions of the term, 62–63
  meaning and, 61, 62, 90, 107, 141–142
  notions of, 84, 85
  Otlet on, 41
  post-documentation technologies and, 139
  size, 41
Documents: Doctrines, archéologie, beaux-arts, ethnographie (magazine), 47
Domestication and domesticity, 123, 125, 129, 130. *See also under* Animals
Doty, P., 84
Dreams
  Freud on, 101, 105
  jokes and, 101
Drucker, Johanna, 97–98
Duty rights. *See* Obligation rights

Ecological bodies. *See* Bodies
"Ecology of selves," 127
Ecosystems, 128, 131, 132. *See also* Bodies
  rights and, 131, 132
Emotions, 65, 79–86. *See also* Affect(s); Feelings
Empathy
  animals and, 122, 126, 132, 133

Index

rights and, 126, 132, 133
Empirical particulars, 2, 51, 53, 90
Empirical sciences, 66, 137. *See also*
  Empiricism: in the sciences
Empiricism, 39, 56
  ontology and, 56, 140, 151
  of post-documentation technologies,
    137, 140
  in the sciences, 140, 151 (*see also*
    Empirical sciences)
Enframing (*Gestell*), 14–15
Engineering and science, 15–16, 18
Enlightenment, Age of, 111, 116, 131
Enlightenment (spiritual), 59
Environmental bodies. *See* Bodies
"Environmental semiotics," 126–127
Epistemology, 33, 37–39, 60, 143,
    154n6. *See also* Empiricism
  Bataille and, 38, 47, 51
  documentarity and, 1, 2, 26, 37
  information and, 2, 3, 41
  Latour and, 8, 21–23, 26, 30
  ontology and, 1, 23, 26, 30, 120
  Otlet and, 38, 40, 41, 44, 46–47, 51
  positivist, 22, 37, 44, 61 (*see also*
    Positivism)
  Wittgenstein and, 44
Essence, 51, 119
  animals and, 3, 123, 129
  entities and, 3–5, 14, 15, 20, 21
  Heidegger and, 14–15, 18, 20, 122
  rights discourse and, 131
  strong and weak documentarity and,
    20, 34
  transcendental, 46, 94, 131
  universal, 51, 53, 59, 64
Ethics, 44. *See also* Morality; Rights
  animal, 124–126 (*see also* Animal
    rights)
Ethnology, 37–39
Evidence. *See also specific topics*
  document as, 84
  inscription and, 151, 152
  modes of, 2, 11, 85, 111, 157n2

ontology and, 1, 12
  philosophy of, 1, 2, 84–85 (*see also*
    Documentarity)
Existence. *See also* Being; *Dasein*
  modes of, 122, 123
Expression, 9, 99, 115, 138. *See also specific topics*
  dispositional-affordance theory of,
    31
  of dispositions, 30, 130, 137
  Latour on, 22–25, 27
  modes/modalities of, 20, 22–24, 52,
    129, 138 (*see also* Latour, Bruno: *An
    Inquiry into Modes of Existence*)
  ontology and, 30, 53, 110
  rights of, 65, 111, 112, 123, 130, 131,
    133 (*see also* Agency rights; Bodies:
    rights of; Rights)
  trauma and, 105–107
Expressions
  affordances and, 137 (*see also*
    Affordances)
  dispositions and, 31, 33–35
  of entities, 9, 16, 24, 32–35, 53, 137
  of selves (*see* Self-expression)
  social and cultural, 32, 33 (*see
    also* Cultural affordances; Social
    affordances)
Expressive agency, 127, 129
Expressive particulars, 31, 126
Expressive powers, 2, 39, 69, 126, 133
  Harré's theory of, 30–35
Expressive rights. *See* Expression: rights
  of

Fables, 106–108
Fairy tales, 99, 106–107
Fantasia. See Imagination
Feelings, 65, 66, 93, 121. *See also*
  Affect(s); Emotions
Ferraris, Maurizio, 8–9, 109, 152
Fiction. *See* Literature; Realist fiction/literary realism
Figural interpretations, 55

Figuration
  calculation and, 63–64
  indexicality and, 61, 63
  and the indexical sign (*indice*), 56–60
  religious, 57
  rhetorical, 52, 55, 57, 60–61, 63
  role in philosophical historicism, 61
Flaubert, Gustave. *See Madame Bovary*
Formalism, 2, 85, 89–90
  constructivism and, 85, 89, 90
Foucault, Michel, 21
Freedom of Information Act (FOIA), 112
Freitas, Lídia, 157n1
Freud, Sigmund, 100–101, 104
Frohmann, Bernd, 154n6

Gagliano, M., 132
Garfield, Eugene, 142
Gender roles, 71
  in *Madame Bovary*, 80–82
Generalization (machine learning), 150
Gibson, James J., 30–32
"Go along with" humans, 124
  androids' inability to, 158n4
  animals unable to, 123, 124
  Heidegger on the concept of, 122–125
  refusal to, 123
  (rights for) animals who can, 122–125, 131
  rights to not, 130
  "world" and, 123
God. *See* Ontotheology of the West; Religion
Google PageRank, 140–142, 159n3
Google Search, 142, 143. *See also* Google PageRank
Grammar in jokes, 100–103
Guattari, F., 123

Harré, Rom, 9, 30–32
Harryman, Carla, 85, 91, 94–97
Hatred, jokes and, 103–104

Hegel, Georg Wilhelm Friedrich, 61
Heidegger, Martin, 21, 38, 57, 121–126, 158n4
  affordances and, 14–15, 18, 32
  on *aition* (causes/causal affordances), 18, 32
  animals, the world, and, 121–126, 158n4
  on concept of "go along with" humans, 122–125
  on concern/care, 122
  on cybernetics, 16, 17
  on *Dasein*, 122, 123, 125
  essence and, 14–15, 18, 20, 122
  Husserl and, 154n4
  on information theory, 17
  language and, 17–20, 32, 159n4
  materials and, 18
  metaphysics and, 12–15, 17, 20, 122, 124
  modernity and, 12–14, 20
  ontology and, 12, 13, 19, 20, 57, 123–125
  poiesis and the task of thinking, 11–20 (*see also Poiesis*: Heidegger and)
  science and, 12–16, 19
  techne and, 9, 13–15, 18, 20, 158n3
  on technology, 16
  on transposability, 122–123
  on "world-picture," 14, 17
  writings of
    Being and Time, 11–14, 19–20, 122, 154n4
    "The End of Philosophy and the Task of Thinking," 11–13, 15, 16, 19
Hermeneutics, 55, 58
"High style" literature, 66–67, 70, 80
High style (rhetorical style), 53
"High style" sense of classicism, 55
History, philosophy of, 61
The Human Comedy (Balzac), 70–72
Human rights, 112, 114, 116–119. *See also* Natural rights; Rights

Index

animals and, 118–119, 125, 130, 131, 133 (*see also* Animal rights)
entities and, 125

Icons and iconography, 57–59
Christian, 57–60
Kohn on, 127, 128
medieval, 56–60
Walsh on, 56–60
Ideal reference, 2, 5, 38. *See also* Strong documentarity
Imagination (*fantasia*), 73, 121, 122, 129
animals and, 121, 122, 129
Aristotle on, 120
Heidegger and, 14
(lower vs. higher) levels of, 120, 121
representation and, 129, 152, 157n3
Imaginative transposition, 122. *See also* Transposability
Immutable mobiles, 26, 27
Imperialism, 116–117
Indexicality, 27, 126–129
animals and, 126, 129
figuration and, 61, 63
Kohn on, 127–129
language and, 57, 102–103
ontology and, 27, 61, 129
semiotics and, 126, 129
Indexical points, 29, 49, 61, 127
Indexical reference, 127, 129
Indexical relationships, 126, 128
Indexical sign (*indice*), 56, 108, 128
animal entity as, 129
documentation theory and the, 60–64
figuration and the, 56–60
Indexing It All: The Subject in the Age of Documentation, Information, and Data (Day), 3, 9, 158n4, 159n1
Indian Protection Service (SPI) in Brazil, 116, 117
Indice. *See* Indexical sign

Indices, 127
Indigenous peoples, 117, 118, 133
Inductive bias, 150
Information. *See also specific topics*
characteristics and qualities of, 2–3
conceptions and notions of, 3, 84, 85, 109
defining, 28
documentation and, 2, 68
inscription and, 28
metaphysics and, 3, 17
modern human rights related to, 111–112
modernity and, 65, 85, 109, 142, 143, 148, 158n1
"Otlet" vs. "Cutter" tradition of, 40–41
from poetics to a critique of knowledge and, 97–98
poiesis as, 29
Information access, 115
Information access rights, 112, 130
Information age, 145–146, 159n1
Informational fragments, 108–109
Information infrastructures, 98, 138
Information science. *See* Library and information science
Information theory, 17, 92
Inscription, 118, 149. *See also specific topics*
connotations of the term, 22
documentarity and, 116
evidence and, 151, 152
information and, 28
modes of, 1, 20, 30, 52, 151
"nonsense" inscriptions, 87, 87f
ontology and, 20–23, 30
representation and, 150, 152
science and, 21, 22, 27, 38
self and, 118
technologies and, 125, 137, 138, 149
Insults, jokes and, 101–103
Intellectual freedom, 112

Jokes, 100–104
  grammar in, 100–103
  hatred and, 103–104
  insults and, 101–103
  language and, 100–103
  meaning and, 99–102
  rhetoric and, 102–104
  tendentious vs. non-tendentious, 101
Judgment, technologies of, 1, 8, 9, 35
Judgments, 9
  aesthetic, 149, 152
  computer-mediated, 137–138, 148–149
  moral, 35, 52, 121 (*see also* Morality)

Kant, Immanuel, 87, 111, 122
"Karawane" (Ball), 87, 87f
Kichwa. *See* Runa people in Ecuadorian Amazon
Knowledge
  "coloniality of knowledge," 116
  conceptions of and nature of, 19, 142
  documentarity and forms of, 37
  modernity and, 74, 109, 142, 143, 148
  Otlet on, 40–47
  from poetics to a critique of information and, 97–98
"Knowledge organization" classification systems. *See* Library of Congress Subject Headings
Knowledge processes, role of representation in, 155n2
Kohn, Eduardo, 126–129, 133, 158n5
  on animal rights and rights of nature, 132, 133, 158n5
  on animals, 128–129
  cybernetics and, 132, 133
  on indexicality, 127–129
  on language, 128, 129
  ontology and, 129, 130
  on self, 127, 128
  on semiotics, 126–130, 133

Language, 88, 118. *See also* Library of Congress Subject Headings; *specific topics*
  animals and, 119, 121
  cybernetics and, 16, 17
  Heidegger and, 17–20, 32, 159n4
  indexicality and, 57, 102–103
  jokes and, 100–103
  "Karawane" (Ball) and, 87, 87f
  Kohn and, 128, 129
  Latour and, 25–27, 57
  The Maintains (Coolidge) and, 88–89
  materiality/material base of, 90
  meaning and, 90, 104, 140, 142, 159–160nn3–4
  natural, 160n4
  philosophy of, 26, 140, 142, 144, 159n3
  post-documentation's, 140–143
  picture theory of, 41
  representation and, 96, 119
  Thomas and, 144
  transparency and, 96
  trauma and, 104
  Walsh and, 58
  Watten and, 89–91
  Wiener and, 16–17
  Wittgenstein and, 26, 41, 140, 142
Language games, 156n1, 159n4
Language Writing/Language Poetry, 67, 89–97
Latour, Bruno, 8, 29–30, 57
  affect and, 23, 24
  affordances and, 8, 21, 22, 25, 27
  on center of calculation (*centre de calcul*), 27–29, 63, 119
  on (chains of) reference, 26, 27
  epistemology and, 8, 21–23, 26, 30
  on expression, 22–25, 27
  An Inquiry into Modes of Existence: An Anthropology of the Moderns, 8, 22, 23

Index

inscription in the works of, 20–30
Kohn and, 128, 151
language and, 25–27, 57
on libraries, 27–29
on maps, 26–29, 81, 119, 128, 151
on material(s), 23–25, 28, 29
on meaning, 25, 27, 29
ontology and, 20–23, 25–27, 30, 57, 62, 119
poiesis and, 20, 21
pragmatism and, 26, 27
on prepositions, 25, 26
science and, 21–23, 27, 29
substance and, 22, 25, 29, 30
Leake, David B., 160n6
Learning algorithms, 141, 145, 149–150. *See also* Algorithms
Leclerc, Georges-Louis (Comte de Buffon), 71–72
Lhermitte, Jean-François, 120–122
Librarians, 38, 46, 61
school, 160n2
Libraries, 27–29, 38, 41, 44
Latour on, 27–29
in *Madame Bovary*, 82–84
nature of, 27–29, 146, 152
Otlet and, 29, 38, 40, 44
representations or "metadata" held by, 43
Library and information science, 37, 113, 119, 151
Library of Congress Subject Headings (LCSH), 141, 143, 159n4
Link-analysis systems, 142, 145
Literature (fiction), 65, 66. *See also* Realist fiction/literary realism; *specific topics*
constructivism and, 85, 86, 89, 90
defined, 65
documentation and, 65, 68
function of, 57
"low"/"lower" style of, 55, 67

modernity and, 65, 66, 68, 74, 81–82, 85
nature of, 65, 66
ontology and, 67, 72, 85
questions about the relationship between information and, 68
science and, 47, 51, 66, 68–71, 78, 81, 82, 84
weak documentarity and, 53, 69, 70, 85
Locke, John, 111–112
Logos, 46, 64
"Low style" literature, 53, 67

Machine learning, 148–150
Madame Bovary (Flaubert), 69–86, 115, 156n2
aristocracy in, 73, 80, 81, 84
bourgeoisie in, 70, 72–74, 78–81, 84, 115, 156n2
gender roles in, 80–82
libraries in, 82–84
morality and, 69–70, 73, 75, 78, 80–82, 84, 115
sexuality in, 80
social class in, 53, 67, 79–81, 84
The Maintains (Coolidge), 88–89
Maps, Latour on, 26–29, 81, 119, 128, 151
"Material" affordances. *See* Affordances
Materialism, 37, 66
Bataille and the philosophy of base materialism, 38, 45–47, 49, 51
Latour on, 23 (*see also under* Latour, Bruno)
Materiality, 32, 42, 86, 90
McCraken, Peggy, 157n2
McNeill, W., 121–122
Meaning, 51, 52, 57, 58, 61, 87f, 88, 140–142, 145
avant-garde and, 88, 96
Briet and, 61, 62

Meaning (cont.)
  constructivism and, 90 (*see also* Constructivism)
  creation of, 29, 55, 92, 93, 95, 96, 144, 159–160nn3–4 (*see also* Constructivism)
  cultural and social, 32
  documents and, 61, 62, 90, 107, 141–142
  expressions and, 99, 138
  jokes and, 99–102
  Kohn on, 128
  language and, 90, 104, 140, 142, 159–160nn3–4
  Latour on, 25, 27, 29
  psychoanalysis, the unconscious, and, 100–101, 104
  reference for, 138
  of representations, 95
  revelation and, 51, 57, 59, 60, 64
  sense and, 25
  symbol as basis for, 66
  texts as potential spaces for, 145
  theology and, 51, 55, 57–60
  trauma and, 104, 107
Metadata, 41–43, 143
Metaphysical ontology, 13, 15, 118. *See also* Metaphysics: ontology and
Metaphysical traditions, 121, 124, 153n1. *See also* Metaphysics: Western
Metaphysics, 90, 129
  Bataille and, 49, 51
  documentarity and, 1, 20, 38, 49, 68, 116–119
  Heidegger and, 12–15, 17, 20, 122, 124
  of information, 3
  information as, 3
  information exchange and, 17
  ontology and, 12, 13, 15, 20
  Western, 1, 12, 13, 64, 84, 116, 119, 146, 153n1
  Mignolo, Walter D., 116–117

Mimesis: The Representations of Reality in Western Literature (Auerbach), 53–56. *See also* Auerbach, Erich
Minimalism, 88
Mitsein (being-with), 122, 123. *See also* Being-with
Modern documentary tradition, 81, 159n1
Modern documentation theory, 52. *See also* Documentation theory
Modern documentation tradition, 3
The Modern Invention of Information: Discourse, History, and Power (Day), 3, 145, 147, 154n3
Modernity, 14, 21, 45, 109, 117
  documentarity and, 65, 68, 74
  Heidegger and, 12–14, 20
  information and, 65, 85, 109, 142, 143, 148, 158n1
  knowledge and, 74, 109, 142, 143, 148
  literature and, 65, 66, 68, 74, 81–82, 85
  and the media, 148, 157n4
  Mignolo on, 116, 117
  nature of, 13
Monographic principle (Otlet), 41
Moral character, 69–70, 74, 154n7
Morality, 35, 74, 120, 121, 132, 154n7. *See also* Ethics; Judgment: moral
  Madame Bovary and, 69–70, 73, 75, 78, 80–82, 84, 115
  systems science and, 15
Moral orders, 72, 73
Moral perfection, 120, 121d
Moral rights, 119. *See also* Rights
Museums, 38–39
Mystery plays, 59

Napo Kichwa. *See* Runa people in Ecuadorian Amazon
Naqvi, Y., 113–115
Natural bodies. *See* Bodies
Natural entities. *See* Bodies
Naturalism and realism, 67, 69, 70, 90

Natural philosophy, 21–22, 70–72
Natural rights, 111–112, 118, 126, 132, 133
Natural sciences, 24, 32, 33, 130, 137, 150, 155n2
Nature. *See also* Plants
  rights of, 118–119, 125, 130–134
New media, 108–109, 144, 147–148
Nietzsche, Friedrich Wilhelm, 38

Objectivism, metaphysical, 90
Obligation rights, 111, 119, 130, 131
Ontological categories, 25, 120, 125
Ontological naming, 64
Ontological particulars, 30, 67, 110
Ontological rights for entities, 130
Ontological substances, 30. *See also* Ontology: substance and
Ontologists, 141, 142
Ontology
  animals and, 62, 120, 123–126, 129–131, 134
  avant-garde and, 85
  Briet and, 57, 61–64, 119, 129
  defined, 12
  dispositions and, 30, 35
  documentarity and, 1, 20, 21, 26, 67, 118
  documentary, 1
  documentary notions of, 63
  documentary techniques of, 62
  documentation and, 49, 64, 129
  empiricism and, 56, 140, 151
  entities and, 9, 64, 119
  epistemology and, 1, 23, 26, 30, 120
  evidence and, 1, 12
  existential (*see Dasein*)
  expression and, 30, 53, 110
  fiction and, 67, 72, 85
  fundamental, 131
    seeking a, 13, 19–20
  Heidegger and, 12, 13, 19, 20, 57, 123–125
  indexicality and, 27, 61, 129
  information and, 1, 140
  inscription and, 20–23, 30
  Kohn and, 129, 130
  Latour and, 20–23, 25–27, 30, 57, 62, 119
  metaphysical, 13, 15, 118
  metaphysics and, 12, 13, 15, 20
  modernity and, 13
  philosophical, 1
  philosophy and, 1, 12, 63
  poiesis and, 13, 19–20, 23
  representational, 124
  representation and, 1, 13, 27, 43, 119
  rights and, 118, 119, 124–126, 130, 131
  science and, 19, 21, 23, 35, 49, 62, 64, 119, 134, 140, 150
  self and, 72, 74, 125
  semiotics and, 85, 126, 129, 130
  "social," 85
  substance and, 25, 30, 35
  technology and, 1, 13
  typology/taxonomy/classification and, 21, 43, 61, 62, 64, 119, 125, 134, 141 (*see also* Ontologists)
Ontotheological foundation, 57
Ontotheology of the West, 51, 59, 64
Organic bodies. *See* Bodies
Otlet, Paul, 45f, 47–49, 51, 61, 64, 143
  Bataille and, 38, 40, 45–49, 51
  bibliographic positivism, 38, 40–45, 61
  Briet and, 61, 64
  "Cutter" vs. "Otlet" tradition of information, 40–41
  on documentation, 42–43
  epistemology and, 38, 40, 41, 44, 46–47, 51
  on knowledge, 40–47
  libraries and, 29, 38, 40, 44
  overview, 38, 40
  science and, 40–42, 43f, 44

PageRank (Google Search algorithm), 140–142, 159n3
Pain, 131, 132
Paradigms, 154n2
Particulars. *See also* Powerful particulars
  dispositional powers and, 2, 150
  empirical, 2, 51, 53, 90
  expressive, 31, 126
  ontological, 30, 67, 110
  paradigms and, 154n2
  radical, 90–92
Peirce, Charles Sanders, 126–128
Pelizzon, A., 132
Phenomenological sense(s), 2, 5, 143. *See also* Weak documentarity
Philosophical historicism, 61
Philosophy. *See also specific topics*
  final stage and end of, 11–13, 15, 16, 19
Physical/biological affordances, 21, 27, 32, 33. *See also* Affordances
Plants, 71, 129
  rights of, 132, 133 (*see also* Nature: rights of)
  self and, 128
Plato, 121–122
Poetic devices, 68, 89
Poetics, 90, 97, 98
  knowledge and, 96
  Questions of Poetics: Language Writing and Consequences (Watten), 91
Poetry, 17–21, 66, 90, 91. *See also* Language Writing/Language Poetry
Poets, surrealist, 48
Poiesis, 89
  avant-garde and, 96
  Heidegger and, 9, 13, 14, 18–20, 29 (*see also under* Heidegger, Martin)
  as information, 29
  Latour and, 20, 21
  natural vs. human, 18
  ontology and, 13, 19–20, 23
  techne and, 9, 13, 14, 18, 20

Positivism. *See also* Epistemology: positivist
  bibliographic positivism of Otlet, 38, 40–45, 61
  documentarity and, 37, 40
  entities and, 22
  epistemology and, 22, 37, 38, 61
  Heidegger and, 20–21
  Latour and, 22
  science and, 21
  Wittgenstein and, 41, 44
Positivist representation, 37
Post-documentation, 7–8
  philosophy of language, 140–143
  principles, 138–139
Post-documentation perspective, 137
Post-documentation technologies, 139, 141, 144–145, 159n4. *See also* Post-documentation
  empiricism of, 137, 140
  key to understanding, 144
  nature of, 135
Power. *See also* Dispositional powers; Expressive powers
  definitions, 31
  types of, 31
Powerful particulars. *See also* Post-documentation
  dispositions and, 30, 31
  documentarity and, 137
  expression and, 30, 31, 54
  Harré on, 31
  literature and, 54, 55, 65–67, 70, 84, 85
  narratives and, 111
  nature of, 31
  representation and, 134–135
  rights and, 111, 117–118, 131, 133, 134
Prepositions
  Harryman on, 96
  Latour on, 25, 26
  nature of, 25, 96

Presence, 3, 13–15
Progress (Watten), 91–92
Psychoanalysis and trauma, 99, 104–106
Psychology, 21

Questions and questioning, 11–13, 92–93
Questions of Poetics: Language Writing and Consequences (Watten), 91

Radical particulars, 90–92
Rayward, W. Boyd, 40
Real, the, 46, 51, 61, 81, 94, 104, 105, 107, 108
  information age and, 145–146
  inscriptions and, 38
Realism. *See also* Objectivism; Realist fiction/literary realism
  in documentary field research products, 38
  inscription and, 38, 86
  institutional pragmatism and, 30
Realist fiction/literary realism, 2, 66, 67, 69–70, 94, 95, 140. *See also Madame Bovary*
  allegory and, 106–107
  "as-if" structures and, 106–108
  Auerbach's history of, 53–56
  avant-garde and, 66, 85–86, 96–97
  character(s) and self(ves) in, 69–70, 72
  constructivism and, 67, 89–90
  fables compared with, 108
  fairy tales, folktales, and fables, and, 106–108
  modern, 51–55, 70, 84–85
  naturalism and, 67, 69, 70, 90
  nature of, 55, 69, 94, 96–97
  "realistic" traditions in, 96–97
  representational, 85
  representation and, 66, 67, 85–86, 90, 96–97, 106
  social sciences and, 69
  surrealism and, 46
Realist genres, 140

Reasoning, modes of, 156n1
Reddy, Michael J., 92
Reductionism, 33
Reference, 138. *See also* Ideal reference; Strong documentarity
Religion. *See also* Ontotheology of the West; Theology
  medieval religious iconography as indexical signs, 56–60
  religious figuration, 57
  religious symbolism, 62
  religious texts, 49 (*see also* Biblical narratives)
Representation
  animals and, 119–121, 157n3
  history of, 56
  imagination and, 129, 152, 157n3
  metaphysics of, 15
  modes of, 20, 21, 52, 121, 130, 155n2
  ontology and, 1, 13, 27, 43, 119, 124
  realist fiction/literary realism and, 66, 67, 85–86, 90, 96–97, 106
  science and, 29, 38–40, 42, 52, 155n2
  strong and weak documentarity and, 2, 20, 38, 39, 56
Representational imagination, 152
Reproduction, 23–25
Revelation, 51, 58–60
  documentarity and, 49, 56, 60
  entities and, 64
  indexical points of, 49
  meaning and, 51, 57, 59, 60, 64
  theology and, 51, 56, 59, 60
  truth and, 49, 56–57, 60, 64
Revelatory modes, 52. *See also* Realist fiction/literary realism; Representation: modes of
  ontological and historiographic, 57
Revelatory narratives, 56–57
Revelatory understanding of scientific documentation, 60. *See also* Indexical sign: documentation theory and the

Revelatory unfolding, 59. *See also* Revelation
Rhetoric
jokes, insults, and, 102–104
spirit of, 54
Rhetorical devices, 56, 58, 95, 107, 108, 156n1
Rhetorical figuration, 52, 55, 57, 60–61, 63
Rhetorical forms, 103–104, 108, 147
Rhetorical genres, 14, 63
Rhetorical performances, 14, 44
Rhetorical strategies, 51, 58, 59, 104
Rhetorical styles, 53
Rhetorical "switches," 102–103
Rhetorical units, 44, 89
Rights. *See also* Animal rights; Bodies: rights of; Expression: rights of
direct vs. indirect, 112
dispositions and, 111, 130, 133
empathy and, 126, 132, 133
ontology and, 118, 119, 124–126, 130, 131
"Rights drift," 130–131
Rimbaud, Arthur, 61, 66, 138
Runa people in Ecuadorian Amazon, 126, 127
Russian formalism, 89–90

Šabanović, Selma, 157n2
Safian, Louis A., 102
Scholarship, 139, 152
Science, 21, 30, 33, 134, 139, 152. *See also* Empiricism; *specific sciences*
affordances and, 32, 35, 137
Briet and, 63, 64
conceptions of and nature of, 14, 19, 21, 30, 37–38, 40, 41, 49, 64, 139
cybernetics and, 16
Debaene and, 38, 39
dispositions and, 21, 32–33, 137
documentarity and, 39, 49, 56, 57
documentation and, 49, 52, 53, 56, 60–64
empirical sciences, 66, 137
empiricism in, 140, 151
engineering and, 15–16, 18
entities and, 22, 32, 33, 35, 64
Heidegger and, 12–16, 19
inscription and, 21, 22, 27, 38
Latour and, 21–23, 27, 29
literature and, 47, 51, 66, 68–71, 78, 81, 82, 84
"man of science" (Bayer), 158n1
modes of, 21–22, 39
ontology and, 19, 21, 23, 35, 49, 62, 64, 119, 134, 140, 150
Otlet and, 40–42, 43f, 44
philosophy and, 16
poiesis and, 21
representation and, 29, 38–40, 42, 52, 155n2
rights and, 110, 115, 134
Science Citation Index, 142, 143
Scientific research, 15, 18, 19, 22, 134
Scientific revelation, 49, 64. *See also* Science
Scopsi, Claire, 159n1
Scott, Walter, 72
Scudéry, Madeleine de, 81, 82f
Secondhand knowledge, 65, 74, 78, 79, 158n1. *See also* Documentary knowledge
Self, 35
affects and, 127, 128
conception of (individual), 65, 66
defined, 127
dispositions and the, 137, 155n7
"ecology of selves," 127
in literature, 69–70
nature of, 66, 127, 155n7, 156n2
semiotics and, 127, 128
Self-expression, 53, 127, 131, 134, 153n2

Semiotic being, 126–130
Semiotic indexicality, 129
Semiotic narratives, 133
Semiotic notion of life (system), 132
Semiotics, 58
  affect and, 128, 129
  animals and, 126, 129, 133
  "environmental," 126–127
  indexicality and, 126, 129
  Kohn and, 127–130
  and the natural world, 128–130, 132, 133
  ontology and, 85, 126, 129, 130
  Peirce's, 126–128
  self and, 127, 128
Sense, 20, 138, 143–146. *See also* Animals: sensibility/perception; Phenomenological sense(s); Weak documentarity
  affects and, 20, 24–25
  definition and nature of, 24–25, 52, 72
  of documents, 114
  and reference in the media square, 146–148
Sexuality, 46, 51, 80
Signification, condensation and displacement of, 101
Signs, 57–59. *See also* Indexical sign
Slavery, 118
Social affordances, 7, 21, 32, 33. *See also* Affordances
Social class, 77–81, 84, 118. *See also* Aristocracy; Bourgeoisie; "High style" literature; "Low style" literature
  in *Madame Bovary*, 53, 67, 79–81, 84 (*see also Madame Bovary*)
Social facts, 153n2
Social networks, 22, 134, 140, 141, 143–144
Social sciences, 33, 69, 70
  fictional realism and, 69, 70, 155n2
Social sense, 142, 144, 148

Sterne, Laurence, 66
Strong documentarity, 21, 34, 37–40, 52, 56, 67, 137. *See also under* Weak documentarity
  critiques of, 20, 26
  definition and nature of, 2, 61
  Harryman and, 85, 94
  Heidegger on, 20
  Latour on, 26
  metaphysics of, 20, 38, 49
  modern philosophy and, 68
  ontology and, 21, 26
  realism and, 70
  representation and, 38, 39
  rights and, 130
Strong reference, 33–34. *See also* Strong documentarity
"Strong" semantic techniques of classification, 85
Structured data, 145
Substance, 34–35
  dispositions and, 34, 35
  entities and, 25, 34–35
  expression and, 34
  inscription and, 30, 34
  Latour and, 22, 25, 29, 30
  notions of, 35
  ontology and, 25, 30
  ontology of, 35
  philosophy of, 29, 30
  theory of, 30
Substances, theory of, 30
Suffering, 131, 132
Surrealism, 46, 48

Tagging systems, 159n4
Taxonomy(ies), 21, 26, 29, 61, 62, 64. *See also under* Ontology
  animals and, 28–29, 62, 125, 130, 134
  rights and, 125, 130
Techne, 119, 138, 158n3. *See also* Technology(ies)

Heidegger and, 9, 13–15, 18, 20, 158n3
poiesis and, 9, 13, 14, 18, 20
Technological transfer, 157–158n3
Technology(ies), 16. *See also* Techne
  inscription and, 125, 137, 138, 149
  ontology and, 1, 13
Technoscience, 16, 19, 137
Theology. *See also* Ontotheology of the West; Religion
  meaning and, 51, 55, 57–60
  revelation and, 51, 56, 59, 60
Thinking, Heidegger on *poiesis* and the task of, 11–13, 16, 20
Thomas, Neal, 144–145
Transcendental essence, 46, 94, 131
Transfiguration, 64
Transmutation, 62
Transposability, Heidegger on, 122–123
Transubstantiation, 64
Trauma, 104–106
  psychoanalysis and, 99, 104–106
Trees, 132. *See also* Nature: rights of; Plants
Truth, 118
  right to, 112–118
Truth and reconciliation commissions, 113, 114, 118
Typology, 55, 56

Unconscious, 100–101
Universal essence, 51, 53, 59, 64
Universal-particular framework, 154n2

Virtual, the, 145

Walsh, John A., 52, 56–61
Watten, Barrett, 89–94, 97
  language and, 89–91
Weak documentary, 20, 38, 39, 72
  built on a method of strong sense, 52
  historical progress from strong documentarity to, 56
  literature and, 53, 69, 70, 85
  representation and, 2, 20, 38, 39, 56
  shift from strong documentarity to, 2, 53, 56, 69, 70, 137
  strong documentarity and, 20, 39, 72, 85
  strong documentarity coming in the back door of, 150
Weaker sense, 34. *See also* Weak documentarity
What Is Documentation? (Briet), 3–5, 60–64, 125, 129, 159n1
Wiener, Norbert, 16
Willson, M., 40–41
Wittgenstein, Ludwig, 41, 44
  language and, 26, 41, 140, 142
"World," 59, 124
  conceptions of, 122, 123
  Heidegger on, 122, 123
"World-picture," 14, 17
Writing
  Language Writing/Language Poetry, 67, 89–97
  as social contestation and construction, 89–90

Zola, Emile, 70